D1544765

Paul Feyerabend

NLB

Science in a Free Society

First published by NLB 1978

© Paul Feyerabend 1978

NLB, 7 Carlisle Street, London W1
Filmset by Servis Filmsetting Ltd, Manchester

Printed in Great Britain by Lowe & Brydone Ltd,
Thetford, Norfolk

Designed by Ruth Prentice

ISBN 86091 008 3

Contents

Preface

The essays in this volume resume the argument I started in *Against Method* (*AM* for short) and develop it further. There are replies to criticism, there is new material I prepared for the paperback edition but could not use, and there is an extended discussion of relativism and the role of science (rationalism) in a free society. Like the earlier book this volume has one aim: to remove obstacles intellectuals and specialists create for traditions different from their own and to prepare the removal of the specialists (scientists) themselves from the life centres of society.

Parts One and Two have one single aim: to show that rationality is one tradition among many rather than a standard to which traditions must conform. Part One develops the argument for science, Part Two extends it to society as a whole. In both cases the basic theoretical problem is the *relation between Reason and Practice. Idealism* assumes that Practice (the practice of science, of art, speaking a natural language, custom as opposed to formal laws) is crude material to be formed by Reason. Practice may contain elements of Reason, but in an accidental and unsystematic fashion. It is the conscious and systematic application of Reason to a partly structured, partly amorphous material that gives us Science, a Society worth living in, a History that can pride itself on having been made by men at their best.

Naturalism, on the other hand, assumes that history, the law, science are already as perfect as they can be. Men do not act without thought and they always try to reason as well as they can. The results are imperfect partly because of adverse conditions, partly because good ideas do not arrive before they do. The attempt to rearrange science or society with some explicit theories of rationality in mind would disturb the delicate balance of thought, emotion, imagination and the historical conditions under which they are applied and would create chaos, not perfection. This was Herder's (and Hamann's) criticism of the enlightenment, this point was seen by Lessing despite his rationalist bias, this was Burke's objection to those who wanted to reform society with the help of well-constructed blueprints, this is the objection raised again by Polányi, Kuhn and others against idealistic philosophies of science. To understand all the many possibilities of Reason, says the naturalist, one has to see it in action, one has to analyse history and its temporal products rather than following the anaemic ideas of those unfamiliar with the richness of science, poetry, language, common law and so on.

Idealism and naturalism have disadvantages which are related (they are mirror images of each other) but can be removed by *combining naturalism and idealism and postulating an interaction of Reason and Practice*. Section 2 explains what 'interaction' means and how it works, Sections 3 to 6 provide illustrations from the sciences. Section 3, for example, shows how even the most abstract standards, standards of formal logic included, can be criticized by scientific research, Section 5 resumes the discussion of the so-called 'Copernican Revolution' and shows why it cannot be captured by any rationality theory: the same argument, presenting the same relations between concepts and based on the same well-known assumptions may be accepted, even praised at one time and fall flat on its face at another. Copernicus' claim to have developed a system of the world in which each part is perfectly adapted to all other parts and where nothing can be changed without destroying the whole meant little to those who were convinced that the basic laws of nature appeared in everyday experience and who therefore regarded the clash between Aristotle and Copernicus as a decisive objection against the latter. It meant a lot to mathematicians who distrusted commonsense. It was read with care by astronomers who despised the ignorant Aristotelians of their time and regarded The Philosopher Himself with contempt – no doubt without having read him. What emerges from an analysis of individual reactions to Copernicus is that *an argument becomes effective only if supported by an appropriate attitude and has no effect when the attitude is missing* (and the attitude I am talking about must work *in addition to* the readiness to listen to argument and is independent of an acceptance of the premises of arguments). This *subjective aspect of scientific change* is related to (though never completely explained by) objective properties: every argument involves *cosmological assumptions* which must be believed or else the argument will not seem plausible. *There is no purely formal argument.*

Interactionism means that Reason and Practice enter history on equal terms. Reason is no longer an agency that directs other traditions, it is a tradition in its own right with as much (or as little) claim to the centre of the stage as any other tradition. Being a tradition it is neither good nor bad, it simply is. The same applies to all traditions – they are neither good nor bad, they simply are. They become good or bad (rational/ irrational; pious/impious; advanced/'primitive'; humanitarian/vicious; etc.) only when looked at from the point of view of some other tradition.

'Objectively' there is not much to choose between anti-semitism and humanitarianism. But racism will appear vicious to a humanitarian while humanitarianism will appear vapid to a racist. *Relativism* (in the old and simple sense of Protagoras) gives an adequate account of the situation which thus emerges. Powerful traditions that have means of forcing others to adopt their ways have of course little use for the relational character of value judgements (and the philosophers who defend them are helped by some rather elementary logical mistakes) and they can make their victims forget it as well (this is called 'education'). But let the victims get more power, let them revive their own traditions and the apparent superiority will disappear like a (good or bad – depending on the tradition) dream.

Part Two develops the idea of a free society and defines the role of science (intellectuals) in it. *A free society is a society in which all traditions have equal rights and equal access to the centres of power* (this differs from the customary definition where *individuals* have equal rights of access to positions *defined by a special tradition* – the tradition of Western Science and Rationalism). A tradition receives these rights not because of the importance (the cash value, as it were) it has for outsiders but because it gives meaning to the lives of those who participate in it. But it can also be of interest for outsiders. For example, some forms of tribal medicine may have better ways of diagnosing and treating (mental and physical) illness than the scientific medicine of today and some primitive cosmologies may help us to see predominant views in perspective. To give traditions equality is therefore not only *right* but also *most useful*.

How can a society that gives all traditions equal rights be realized? How can science be removed from the dominant position it now has? What methods, what procedures will be effective, where is the theory to guide these procedures, where is the theory that solves the problems which are bound to arise in our new 'Free Society'? These are some of the questions raised wherever people try to free themselves of restrictions imposed by alien cultures.

The questions assume that there must be *theories* to deal with the problems and they insinuate, ever so gently, that the theories will have to be provided by *specialists*, i.e. *intellectuals*: *intellectuals* determine the structure of society, *intellectuals* explain what is possible and what not, *intellectuals* tell everybody what to do. But in a free society intellectuals are just one tradition. They have no special rights and their views are of no special interest (except, of course, to themselves). Problems are solved not by specialists (though their advice will not be disregarded) but by the people concerned, in accordance with the ideas *they* value and by procedures *they* regard as most appropriate. People in many countries now realize that the law gives them more leeway than they had assumed; they gradually conquer the free space that has so far been occupied by specialists and they try to expand it further. Free societies will arise from such activities, not from ambitious theoretical schemes. Nor is there any need to guide the development by abstract ideas or a philosophy, such as Marxism. Those who participate in it will of course use ideas, different groups will try to learn from each other, they may adapt their views to some common aim and so more unified ideologies may temporarily arise. But such ideologies will come from decisions in concrete and often unforeseeable situations, they will reflect the feelings, the aspirations, the dreams of those making the decisions, they cannot be foreseen by the abstract thoughts of a group of specialists. They will not only reflect what people want and what they are, they will also be *more flexible, better adapted* to particular problems than what sociologists (Marxist, Parsonians etc.), political scientists or just any intellectual may dream up in their offices. This is how the efforts of special groups combining flexibility and respect for all traditions will gradually erode the narrow and self-serving 'rationalism' of those who are now using tax money to destroy the traditions of the taxpayers, to ruin their minds, rape their environment and quite generally to turn living human beings into well-trained slaves of their own barren vision of life.

Part Three contains replies to reviewers whose reactions can be regarded as typical. I have rewritten most of them and I publish them because they develop points only hinted at in *AM*, because even a one-sided debate is more instructive than an essay and because I want to inform the wider public of the astounding illiteracy of some 'professionals'. Reviews in history, classical philology, mathematics, business reviews, review essays such as those published in *Science, Reviews of Modern Physics* or, on a more popular level, in the *Neue Zürcher Zeitung* show competence, intelligence, a firm grasp of the topic discussed and the ability to express difficult matters in simple language. One learns what a school, a book, an article is all about and one is helped to approach it in a critical way. But political philosophy and the philosophy of science have become sinks of illiterate self expression (using forbidding technical terms, of course). Section 3 of Chapter 4 tries to explain why this is so. This section also contains a partial account of the deterioration of philosophy of science from Mach to the Vienna Circle to Popper and his followers.

Part One

Reason and Practice

1. Against Method Revisited

Against Method grew out of lectures I gave at the London School of Economics and University College London. Imre Lakatos attended most of them. His office window at the London School of Economics was directly opposite the window of the lecture hall. He would listen to what I was saying and storm into the lecture hall to raise objections. My aim in the lectures was to show that some very simple and plausible rules and standards which both philosophers and scientists regarded as essential parts of rationality were violated in the course of episodes (Copernican Revolution; triumph of the kinetic theory; rise of quantum theory; and so on) they regarded as equally essential. More specifically, I tried to show (a) that the rules (standards) *were actually violated* and that the more perceptive scientists were aware of the violations; and (b) that they *had to be violated*. Insistence on the rules would not have improved matters, it would have arrested progress.

An argument of this kind makes a variety of assumptions, some of them rather complex. To start with I assume that my readers agree about progress and good science and that they do so independently of whatever rules or standards they adopt. For example I assume that they applaud the gradual acceptance of the idea of the motion of the earth or of the atomic constitution of matter in the late 19th and early 20th centuries independently of what rules and standards they think it obeys. The argument addresses people who hold such beliefs and it tries to convince them that they cannot have both the developments they cherish and the rules and standards they want to defend.

Branch (b) of the argument makes some rather far-reaching assumptions not only about what *did* happen, but about what *could* and what *could not* have happened given the material, intellectual, scientific conditions of a particular time. For example when describing the way in

13

which Galileo separated theory and experience I also point out (*AM*, p. 152) that new correspondence rules not only were not introduced but could not be introduced because it takes time to develop instruments and ways of testing not based on everyday experience. Today Aristotle, tomorrow Helmholtz – that is not only unlikely, it is impossible. Considerations such as these change from one case to the next and so each case must be discussed on its own merits.

In *AM* I discussed two cases in order to create difficulties for Newtonian inductivism, falsificationism and the theory of research programmes. I also tried to show that theories cannot always be compared by content and/or versimilitude, even when they are theories 'in the same domain'. I conjectured that similar difficulties would arise with any rule and any standard not yet completely voided of content. And as rules and standards are usually taken to constitute 'rationality' I inferred that famous episodes in science that are admired by scientists, philosophers and the common folk alike were not 'rational', they did not occur in a 'rational' manner, 'reason' was not the moving force behind them, and they were not judged 'rationally'.

The main objection against such an argument is the poverty of its basis: one or two examples – and rationality is supposed to be done in.[1] Besides, some critics pointed out, the fact that a rule is violated in one case does not make it useless in others, or in the long run. For example, a theory may be in conflict with facts or ad hoc and may still be retained – but *eventually* the conflict will have to be resolved and the ad hoc adaptations will have to be removed.

The reply to the last remark is obvious: changing non adhocness and falsification by facts into non adhocness and falsification *in the long run* means replacing one standard by another and so admitting that the original standard was not adequate. The reply to the first objection is, however, this. It is true that two cases do not all rules remove but as far as I can see they remove basic rules that form an essential part of the rationalists' prayer book. Only some of these basic rules have been discussed in connection with the case studies but the reader can easily apply the assembled material to Bayesean procedures, conventionalism (whether Poincaré or Dingler) and 'conditional rationalism' where rules and standards are asserted to hold under certain well-specified conditions only. He can even remove the demand that scientific research must con-

---

[1] Some readers objected that though I do not seem to mind inconsistencies I still present them as parts of my argument against standard views of rationality. I reply that I assume my readers to be rationalists. If they are not, then there is no need for them to read the book.

form to the laws of logic.[2] Apart from those natural extensions the matter now rests with the rationalist. It is he who suggests that Great Science conforms to Great Standards. What great *and non-empty* standards are to take the place of the standards discussed?

The difficulty of the task is shown very clearly by the fate of the theory of research programmes. Lakatos realized and admitted that the existing standards of rationality, standards of logic included, are too restrictive and would have hindered science had they been applied with determination. He therefore permitted the scientist to violate them (he admits that science is not 'rational' in the sense of *these* standards). However, he demanded that research programmes show certain features *in the long run* – they must be progressive. In Chapter 16 of *AM* (and in my essay 'On the Critique of Scientific Reason'[3]) I have argued that this demand no longer restricts scientific practice. Any development agrees with it. The demand (standard) is *rational*, but it is also *empty*. Rationalism and the demands of reason have become purely verbal in the theory of Lakatos.

It should be noted that I not only *criticize* standards, rules, procedures but also try to show what procedures *aided* scientists in their work. For example, I point out that and why it was reasonable for Einstein to use an unconfirmed and prima facie refuted theory containing internal contradictions in his account of Brownian motion. And I explain why and how the use of a puzzling instrument such as the telescope that was theoretically opaque and showed many unreal phenomena could still contribute to progress. My argument in both cases is *cosmological*: *given* certain properties of the world, of our instruments (theoretical instruments such as standards included) some procedures are bound to fail while others have a chance of succeeding, i.e. of leading to the discovery of details of a world so constituted. For example, I point out that the fluctuations that limit the validity of the strict second law of thermodynamics cannot be identified directly because they occur in all our measuring instruments. Thus *I don't take the excellence of science for granted* (though I often assume it for the purpose of argument), *I try to show wherein it consists and how greatly it differs from the naive standards of excellence proposed by rationalists.*

With this I come to a problem that was never explicitly discussed in *AM* though it underlies all its arguments – the problem of the relation

---

[2] Cf. *AM*, pp. 252ff. and 257f.

[3] Published in C. Howson (ed.) *Method and Appraisal in the Physical Sciences*, Cambridge 1976.

between reason and practice. In *AM* I try to show that reason, at least in the form in which it is defended by logicians, philosophers of science and some scientists does not fit science and could not have contributed to its growth. This is a good argument against those who admire science and are also slaves of reason. They have now to make a choice. They can keep science; they can keep reason; they cannot keep both.

But science is not sacrosanct. The mere fact that it exists, is admired, has results is not sufficient for making it a measure of excellence. Modern science arose from global objections against what went on before and rationalism itself, the idea that there are general rules and standards for conducting our affairs, affairs of knowledge included, arose from global objections to commonsense (example: Xenophanes against Homer). Are we to refrain from engaging in those activities that gave rise to science and rationalism in the first place? Are we to rest content with their results? Are we to assume that everything that happened after Newton (or after von Neumann) is perfection? Or shall we admit that modern science may have basic faults and may be in need of global change? And, having made the admission, how shall we proceed? How shall we localize faults and carry out changes? Don't we need a measure that is independent of science and conflicts with it in order to prepare the change we want to bring about? And will not the rejection of rules and standards that conflict with science forever prevent us from finding such a measure? On the other hand – have not some of the case studies shown that a blunt application of 'rational' procedures would not have given us a better science, or a better world but nothing at all? And how are we to judge the results themselves? Obviously there is no simple way of guiding a practice by rules or of criticizing standards of rationality by a practice.

2. Reason and Practice

The problems I have just sketched are old ones and much more general than the problem of the relation between science and rationality. They occur whenever a rich, well-articulated and familiar practice – a practice of composing, of painting pictures, of stage production, of selecting people for public office, of keeping order and punishing criminals, a practice of worship, of organizing society – is confronted by a practice of a different kind that can interact with it. The *interactions* and their results depend on historical conditions and vary from one case to the next. A powerful tribe invading a country may impose its laws and change the indigenous traditions by force only to be changed itself by the remnants of the subdued culture. A ruler may decide, for reasons of convenience, to use a popular and stabilizing religion as the basic ideology of his empire and may thereby contribute to the transformation both of his empire and of the religion chosen. An individual, repelled by the theatre of his time and in search of something better may study foreign plays, ancient and modern theories of drama and, using the actors of a friendly company to put his ideas into practice, change the theatre of a whole nation. A group of painters, desirous of adding the reputation of being scientists to their already enormous reputation as skilled craftsmen may introduce scientific ingredients such as geometry into painting and thereby create a new style and new problems for painters, sculptors, architects. An astronomer, critical of the difference between classical principles of astronomy and the existing practice and desirous to restore astronomy to its former splendour may find a way to achieve his aim and so initiate the removal of the classical principles themselves.

In all these cases we have a practice, or a tradition, we have certain influences upon it, emerging from another practice or tradition and we observe a change. The change may lead to a slight modification of the original practice, it may eliminate it, it may result in a tradition that barely resembles either of the interacting elements.

Interactions such as those just described are accompanied by changing degrees of *awareness* on part of the participants. Copernicus knew very well what he wanted and so did Constantine the Great (I am now speaking about the initial impulse, not about the transformation that followed). The intrusion of geometry into painting is less easily accounted for in terms of awareness. We have no idea why Giotto tried to achieve a compromise between the surface of the painting and the corporeality of the things painted, especially as pictures were not yet regarded as studies of material reality. We can surmise that Brunelleschi arrived at his construction by a natural extension of the architects' method of representing three-dimensional objects and that his contacts with contemporary scientists were not without consequence. It is still more difficult to understand the gradually rising claims of artisans to make contributions to the same knowledge whose principles were explained at universities in very different terms. Here we have not a critical *study* of alternative traditions as we have in Copernicus, or in Constantine but an *impression* of the uselessness of academic science when compared with the fascinating consequences of the journeys of Columbus, Magellan and their successors. There arose then the idea of an 'America of Knowledge', of an entirely new and as yet unforeseen continent of knowledge that could be discovered, just as the real America had been discovered: by a combination of skill and abstract study. Marxists have been fond of confounding insufficient information concerning the awareness that accompanies such processes with irrelevance and they have ascribed only a secondary role to individual consciousness. In this they were right – but not in the way they thought. For new *ideas*, though often necessary, were not sufficient for explaining the *changes* that now occurred and that depended also on the (often unknown and unrealized) *circumstances* under which the ideas were applied. Revolutions have transformed not only the practices their initiators wanted to change but the very principles by means of which they intended to carry out the change.

Now considering any interaction of traditions we may ask two kinds of questions which I shall call *observer questions* and *participant questions* respectively.

Observer questions are concerned with the details of an interaction. They want to give a historical account of the interaction and, perhaps, formulate laws, or rules of thumb, that apply to all interactions. Hegel's triad: position, negation, synthesis (negation of the negation) is such a rule.

Participant questions deal with the attitude the members of a practice or a tradition are supposed to take towards the (possible) intrusion of another. The observer asks: what happens and what is going to happen? The participant asks: what shall I do? Shall I support the interaction? Shall I oppose it? Or shall I simply forget about it?

In the case of the Copernican Revolution, for example, the observer asks: what impact did Copernicus have on Wittenberg astronomers at about 1560? How did they react to his work? Did they change some of their beliefs and if so, why? Did their change of opinion have an effect on other astronomers, or were they an isolated group, not taken seriously by the rest of the profession?

The questions of a participant are: this is a strange book indeed – should I take it seriously? Should I study it in detail or only superficially or should I simply continue as before? The main theses seem absurd at first sight – but, maybe, there is something in them? How shall I find out? And so on.

It is clear that observer questions must take the questions of the participants into account and participants will also listen most carefully (if they are inclined that way, that is) to what observers have to say on the matter – but the *intention* is different in both cases. Observers want to know what is going on, participants what to do. An observer describes a life he does not lead (except accidentally), a participant wants to arrange his own life and asks himself what attitude to take towards the things that try to influence it.

Participants can be *opportunists* and act in a straightforward and practical way. In the late 16th century many princes became Protestants because this furthered their interests and some of their subjects became Protestants in order to be left in peace. When British Colonial Officials replaced the laws and habits of foreign tribes and cultures by their own 'civilized' laws the latter were often accepted because they were the laws of the King, or because one had no way to oppose them and not because of any intrinsic excellence. The source of their power and 'validity' was clearly understood, both by the officials and by the more astute of their unfortunate subjects. In the sciences and especially in pure mathematics one often pursues a particular line of research not because it is regarded as intrinsically perfect, but because one wants to see where it leads. I shall call the philosophy underlying such an attitude of a participant a *pragmatic philosophy*.

A pragmatic philosophy can flourish only if the traditions to be judged and the developments to be influenced are seen as temporary makeshifts and not as lasting constituents of thought and action. A participant with a pragmatic philosophy views practices and traditions much as a traveller views different countries. Each country has features he likes and things he abhors. In deciding to settle down a traveller will have to compare climate, landscape, language, temperament of the inhabitants, possibilities of change, privacy, looks of male and female population, theatre, opportunities for advancement, quality of vices and so on. He will also remember that his initial demands and expectations may not be very sensible and so permit the process of choice to affect and change his 'nature' as well which, after all, is just another (and minor) practice or tradition entering the process. So a pragmatist must be both a participant and an observer even in those extreme cases where he decides to live in accordance with his momentary whims entirely.

Few individuals and groups are pragmatists in the sense just described and one can see why: it is very difficult to see one's own most cherished ideas in perspective, as parts of a changing and, perhaps, absurd tradition. Moreover, this inability not only *exists*, it is also *encouraged* as an attitude proper to those engaged in the study and the improvement of man, society, knowledge. Hardly any religion has ever presented itself just as something worth trying. The claim is much stronger: the religion is the truth, everything else is error and those who know it, understand it but still reject it are rotten to the core (or hopeless idiots).

Two elements are contained in such a claim. First, one distinguishes between traditions, practices and other results of individual and/or collective human activity on the one side, and a different domain that may act on the traditions without being one on the other. Secondly, one explains the structure of this special domain in detail. Thus the word of God is powerful and must be obeyed not because the tradition that carries it has much force, but because it is outside all traditions and provides a way of improving them. The word of God can start a tradition, its meaning can be handed on from one generation to the next, but it is itself outside all traditions.

The first element – the belief that some demands are 'objective' and tradition-independent – plays an important role in *rationalism* which is a secularized form of the belief in the power of the word of God. And this is how the opposition reason-practice obtains its polemical sting. For the two agencies are not seen as two practices which, while perhaps of unequal value are yet both imperfect and changing human products but as one such product on the one side and lasting measures of excellence on the other. Early Greek rationalism already contains this version of the conflict. Let us examine what circumstances, assumptions, procedures – what features of the historical process are responsible for it!

To start with the traditions that oppose each other – Homeric common-sense and the various forms of rationalism that arise in the 6th to 4th centuries – have *different internal structures*.[4] On the one hand we have complex ideas that cannot be easily explained, they 'work' but one does not know how, they are 'adequate', but one does not know why, they apply in special circumstances only, are rich in content but poor in similarities and, therefore, in deductive connections. On the other side there are relatively clear and simple concepts which, having just been introduced, reveal a good deal of their structure and which can be linked in many ways. They are poor in content, but rich in deductive connections. The difference becomes especially striking in the case of mathematics. In geometry, for example, we start with rules of thumb applying to physical objects and their shapes under a great variety of circumstances. Later on it can be *proved* why a given rule applies to a given case – but the proofs make use of new entities that are nowhere found in nature.

In antiquity the relation between the new entities and the familiar

[4] For details see *AM*, Chapter 17.

world of commonsense gave rise to various theories. One of them which one might call *Platonism* assumes that the new entities are real while the entities of commonsense are but their imperfect copies. Another theory, due to the *Sophists*, regards natural objects as real and the objects of mathematics (the objects of 'reason') as simpleminded and unrealistic images of them. These two theories were also applied to the difference between the new and fairly abstract idea of knowledge propagated by Plato (but found already before) and the commonsense knowledge of the time (Plato wisely uses a distorted image of the latter to give substance to the former). Again it was either said that there existed only one true knowledge and that human opinion was but a pale shadow of it or human opinion was regarded as the only substantial knowledge in existence and the abstract knowledge of the philosophers as a useless dream ('I can see horses, Plato' said Antisthenes 'but I nowhere see your ideal horse').

It would be interesting to follow this ancient conflict through history down to the present. One would learn that the conflict turns up in many places and many shapes. Two examples must suffice to illustrate the great variety of its manifestations.

When Gottsched wanted to reform the German theatre he looked for plays worth imitating. That is, he looked for traditions more orderly, more dignified, more respectable than what he found on the stage of his time. He was attracted by the French theatre and here mainly by Corneille. Being convinced that 'such a complex edifice of poetry (as tragedy) could hardly exist without rules'[5] he looked for the rules and found Aristotle. For him the rules of Aristotle were not a particular way of viewing the theatre, they were the reason for excellence where excellence was found and guides to improvement where improvement seemed necessary. Good theatre was an embodiment of the rules of Aristotle. Lessing gradually prepared a different view. First he restored what he thought to be the real Aristotle as opposed to the Aristotle of Corneille and Gottsched. Next he permitted violations of the letter of Aristotle's rules provided such violations did not lose sight of their aim. And, finally he suggested a different paradigm and emphasized that a mind inventive enough to construct it need not be restricted by rules. If such a mind succeeds in his efforts 'then let us forget the textbook!'.[6]

---

[5] 'Vorrede zum "Sterbenden Cato"' quoted from J. Chr. Gottsched *Schriften zur Literatur* Reclam, Stuttgart 1972, p. 200.

[6] *Hamburger Dramaturgie* Stück 48. Cf., however, Lessing's criticism of the claims of the 'original geniuses' of his time in Stück 96. Lessing's account of the relation between 'reason' and practice is quite complex and in agreement with the view developed further below.

In an entirely different (and much less interesting) domain we have the opposition between those who suggest that languages be constructed and reconstructed in accordance with simple and clear rules and who favourably compare such *ideal languages* with the sloppy and opaque natural idioms and other philosophers who assert that natural languages, being adapted to a wide variety of circumstances could never be adequately replaced by their anaemic logical competitors.

This tendency to view differences in the structure of traditions (complex and opaque vs. simple and clear) as differences in kind (real vs. imperfect realization of it) is reinforced by the fact that the critics of a practice take an observer's position with respect to it but remain participants of the practice that provides them with their objections. Speaking the language and using the standards of this practice they 'discover' limitations, faults, errors when all that really happens is that the two practices – the one that is criticized and the one that does the criticizing – don't fit each other. Many *arguments against* an out-and-out *materialism* are of this kind. They notice that materialism changes the use of 'mental' terms, they illustrate the consequences of the change with amusing absurdities (thoughts having weight and the like) and then they stop. The absurdities show that materialism clashes with our usual ways of speaking about minds, they do not show what is better – materialism or these ways. But taking the participants' point of view with respect to commonsense turns the absurdities into arguments against materialism. It is as if Americans were to object to foreign currencies because they cannot be brought into simple relations (1:1 or 1:10 or 1:100) to the dollar.[7]

The tendency to adopt a participant's view with respect to the position that does the judging and so to create an Archimedian point for criticism is reinforced by certain distinctions that are the pride and joy of armchair philosophers. I refer to the distinction between an evaluation and the fact that an evaluation has been made, a proposal and the fact that the proposal has been accepted, and the related distinction between subjective wishes and objective standards of excellence. When speaking as observers we often say that certain groups accept certain standards, or think highly of these standards, or want us to adopt these standards. Speaking as participants we equally often *use* the standards without any reference to their origin or to the wishes of those using them. We say

[7] Details about the mind-body problem are found in Chapters 9–15 of my essay 'Problems of Empiricism' in *Beyond the Edge of Certainty*, ed. Colodny, New York 1965, preferably in the improved version published in Italian, *I problemi dell' Empirismo*, Milan 1971, pp. 31–69.

'theories ought to be falsifiable and contradiction free' and not 'I want theories to be falsifiable and contradiction free' or 'scientists become very unhappy unless their theories are falsifiable and contradiction free'. Now it is quite correct that statements of the first kind (proposals, rules, standards) (a) contain no reference to the wishes of individual human beings or to the habits of a tribe and (b) cannot be derived from, or contradicted by, statements concerning such wishes, or habits, or any other facts. But that does not make them 'objective' and independent of traditions. To infer from the absence of terms concerning subjects or groups in 'there ought to be . . .' that the demand made is 'objective' would be just as erroneous as to claim 'objectivity' i.e. independence from personal or group idiosyncrasies for optical illusions and mass hallucinations on the grounds that the subject, or the group, nowhere occurs in them. There are many statements that are *formulated* 'objectively' i.e. without reference to traditions or practices but are still *meant to be understood* in relation to a practice. Examples are dates, coordinates, statements concerning the value of a currency, statements of logic (after the discovery of alternative logics), statements of geometry (after the discovery of Non-Euclidean geometries) and so on. The fact that the retort to 'you ought to do X' can be 'that's what *you* think!' shows that the same is true of value statements. And those cases where the reply is not allowed can be easily rectified by using discoveries in value theory that correspond to the discovery of alternative geometries, or alternative logical systems: we confront 'objective' value judgement from different cultures or different practices and ask the objectivist how he is going to resolve the conflict.[8] Reduction to shared principles is not always possible and so we must admit that the demands or the formulae expressing them are incomplete as used and have to be revised. Continued insistence on the 'objectivity' of value judgements however would be as illiterate as continued insistence on the 'absolute' use of the pair 'up–down' after discovery of the spherical shape of the earth. And an argument such as 'it is one thing to utter a demand and quite a different thing to assert that a demand has been made – therefore a multiplicity of cultures does not mean relativism' has much in common with the argument that antipodes cannot exist

[8] In the play *The Ruling Class* (later turned into a somewhat vapid film with Peter O'Toole) two madmen claiming to be God are confronted with each other. This marvellous idea so confuses the playwright that he uses fire and brimstone instead of dialogue to get over the problem. His final solution, however, is quite interesting. The one madman turns into a good, upright, normal British Citizen who plays Jack the Ripper on the side. Did the playwright mean to say that our modern 'objectivists' who have been through the fire of relativism can return to normalcy only if they are permitted to annihilate all disturbing elements?

because they would fall 'down'. Both cases rest on antediluvian concepts (and inadequate distinctions). Small wonder our 'rationalists' are fascinated by them.

With this we have also our answer to (b). It is true that stating a demand and describing a practice may be two different things and that logical connections cannot be established between them. This does not mean that the interaction between demands and practices cannot be treated and evaluated as an interaction of practices. For the difference is due, first to a difference between observer attitude and participant-attitude: one side, the side defending the 'objectivity' of its values *uses* its tradition instead of *examining* it – which does not turn the tradition into something else. And secondly, the difference is due to concepts that have been adapted to such one sidedness. The colonial official who proclaims new laws and a new order in the name of the king has a much better grasp of the situation than the rationalist who recites the mere letter of the law without any reference to the circumstances of its application and who regards this fatal incompleteness as proof of the 'objectivity' of the laws recited.

After this preparation let us now look at what has been called 'the relation between reason and practice'.

Simplifying matters somewhat we can say that there exist three views on the matter.

A. Reason guides practice. Its authority is independent of the authority of practices and traditions and it shapes the practice in accordance with its demands. This we may call the *idealistic version* of the relation.

B. Reason receives both its content and its authority from practice. It describes the way in which practice works and formulates its underlying principles. This version has been called *naturalism* and it has occasionally been attributed to Hegel (though erroneously so).

Both idealism and naturalism have difficulties.

The difficulties of idealism are that the idealist does not only want to 'act rationally' he also wants his rational actions to have results. And he wants these results not only to occur among the idealizations he uses but in the real world he inhabits. For example, he wants real human beings to build up and maintain the society of his dreams, he wants to understand the motions and the nature of real stars and real stones. Though he may advise us to 'put aside (all observation of) the heavens'[9] and to con-

[9] Plato, *Republic* 530af.

centrate on ideas only he eventually returns to nature in order to see to what extent he has grasped its laws.[10] It then often turns out and it often has turned out that acting rationally in the sense preferred by him does not give him the expected results. This conflict between rationality and expectations was one of the main reasons for the constant reform of the canons of rationality and much encouraged naturalism.

But naturalism is not satisfactory either. Having chosen a popular and successful practice the naturalist has the advantage of 'being on the right side' at least for the time being. But a practice may deteriorate; or it may be popular for the wrong reasons. (Much of the popularity of modern scientific medicine is due to the fact that sick people have nowhere else to go and that television, rumours, the technical circus of well equipped hospitals convince them that they could not possibly do better.) Basing standards on a practice and leaving it at that may for ever perpetuate the shortcomings of this practice.

The difficulties of naturalism and idealism have certain elements in common. The inadequacy of standards often becomes clear from the barrenness of the practice they engender, the shortcomings of practices often are very obvious when practices based on different standards flourish. This suggests that reason and practice are not two different kinds of entities but *parts of a single dialectical process*.

The suggestion can be illustrated by the relation between a map and the adventures of the person using it or by the relation between an artisan and his instruments. Originally maps were constructed as images of and guides to reality and so, presumably, was reason. But maps like reason contain idealizations (Hecataeus of Miletus, for example, imposed the general outlines of Anaximander's cosmology on his account of the occupied world and represented continents by geometrical figures). The wanderer uses the map to find his way but he also corrects it as he proceeds, removing old idealizations and introducing new ones. Using the map no matter what will soon get him into trouble. But it is better to have maps than to proceed without them. In the same way, the example says, reason without the guidance of a practice will lead us astray while a practice is vastly improved by the addition of reason.

This account, though better than naturalism and idealism and much more realistic is still not entirely satisfactory. It replaces one sided action (of reason upon practice or practice upon reason) by interaction but it retains (certain aspects of) the old views of the interacting agencies:

[10] *Epinomis*.

reason and practice are still regarded as entities of different kinds. They are both needed but reason can exist without a practice and practice can exist without reason. Shall we accept this account of the matter?

To answer the question we need only remember that the difference between 'reason' and something 'unreasonable' that must be formed by it or can be used to put it in its place arose from turning structural differences of practices into differences of kind. Even the most perfect standards or rules are not independent of the material on which they act (how else could they find a point of attack in it?) and we would hardly understand them or know how to use them were they not well integrated parts of a rather complex and in places quite opaque practice or tradition viz. the language in which the defensor rationis expresses his stern commands.[11] On the other hand even the most disorderly practice is not without its regularities as emerges from our attitude towards non-participants.[12] *What is called 'reason' and 'practice' are therefore two different types of practice* the difference being that the one clearly exhibits some simple and easily producible formal aspects thus making us forget the complex and hardly understood properties that guarantee the simplicity and producibility while the other drowns the formal aspects under a great variety of accidental properties. But complex and implicit reason is still reason and a practice with simple formal features hovering above a pervasive but unnoticed background of linguistic habits is still a practice. Disregarding (or, rather, not even noticing) the sense-giving and application-guaranteeing mechanism in the first case and the implicit regularities in the second a rationalist perceives law and order here and material yet in need of being shaped there. The habit, also commented upon in an earlier part of this section to take a participant's point of view with respect to the former and an observer's attitude towards the latter further separates what is so intimately connected in reality. And so we have finally two agencies, stern and orderly reason on the one side, a malleable but not entirely yielding material on the other and with this all the 'problems of rationality' that have provided philosophers with intellectual (and, let us not forget, also with financial) nourish-

---

[11] This point has been made with great force and with the help of many examples by Wittgenstein (cf. my essay 'Wittgenstein's *Philosophical Investigations*', *Phil. Rev.* 1955). What have rationalists replied? Russell (coldly): 'I don't understand.' Sir Karl Popper (breathlessly): 'He is right, he is right – I don't understand it either!'. In a word: the point is irrelevant because leading rationalists don't understand it. I, on the other hand, would start doubting the intelligence (and perhaps also the intellectual integrity) of rationalists who don't understand (or pretend not to understand) such a simple point.

[12] Cf. my short comments on 'covert classifications' in *AM*, pp. 223f.

ment ever since the 'Rise of Rationalism in the West'. One cannot help noticing that the arguments that are still used to support this magnificent result are indistinguishable from those of the theologian who infers a creator wherever he sees some kind of order: obviously order is not inherent in matter and so must have been imposed from the outside.

The interaction view must therefore be supplemented with a satisfactory account of the interacting agencies. Presented in this way it becomes a triviality. For there is no tradition no matter how hardheaded its scholars and how hardlimbed its warriors that will remain unaffected by what occurs around it. At any rate – what changes, and how, is now either a matter for *historical research* or for *political action* carried out by those who participate in the interacting traditions.

I shall now state the implications of these results in a series of theses with corresponding explanations.

We have seen that rational standards and the arguments supporting them are visible parts of special traditions consisting of clear and explicit principles and an unnoticed and largely unknown but absolutely necessary background of dispositions for action and judgement. The standards become 'objective' measures of excellence when adopted by participants of traditions of this kind. We have then 'objective' rational standards and arguments for their validity. We have further seen that there are other traditions that also lead to judgements though not on the basis of explicit standards and principles. These value judgements have a more 'immediate' character, but they are still evaluations, just like those of the rationalist. In both cases judgements are made by individuals who participate in traditions and use them to separate 'Good' from 'Evil'. We can therefore state:

  i. *Traditions are neither good nor bad, they simply are.* 'Objectively speaking' i.e. independently of participation in a tradition there is not much to choose between humanitarianism and anti-semitism.

  Corollary: rationality is not an arbiter of traditions, it is itself a tradition or an aspect of a tradition. It is therefore neither good nor bad, it simply is.[13]

  ii. *A tradition assumes desirable or undesirable properties only when compared with some tradition* i.e. only when viewed by participants who see the world in terms of its values. The projections of these participants

[13] See *AM*, especially Section 15.

appear objective and statements describing them *sound objective* because the participants and the tradition they project are nowhere mentioned in them. They *are subjective* because they depend on the tradition chosen and on the use the participants make of it. The subjectivity is noticed as soon as participants realize that different traditions give rise to different judgements. They will then have to revise the content of their value statements just as physicists revised the content of even the simplest statement concerning length when it was discovered that length depends on reference systems and just as everybody revised the content of 'down' when it was discovered that the earth is spherical. Those who don't carry out the revision cannot pride themselves on forming a special school of especially astute philosophers who have overcome moral relativism just as those who still cling to absolute lengths cannot pride themselves on forming a special school of especially astute physicists who have overcome relativity. They are just pigheaded, or badly informed, or both.

iii. *i. and ii. imply a relativism of precisely the kind that seems to have been defended by Protagoras.* Protagorean relativism is *reasonable* because it pays attention to the pluralism of traditions and values. And it is *civilized* for it does not assume that one's own village and the strange customs it contains are the navel of the world.

iv. *Every tradition has special ways of gaining followers.* Some traditions reflect about these ways and change them from one group to the next. Others take it for granted that there is only one way of making people accept their views. Depending on the tradition adopted this way will look acceptable, laughable, rational, foolish, or will be pushed aside as 'mere propaganda'. Argument is propaganda for one observer, the essence of human discourse for another.

v. We have seen that individuals or groups participating in the inter-action of traditions may adopt a pragmatic philosophy when judging the events and structures that arise. The principles of their philosophy often emerge only during the interaction (people change while observing change or participating in it and the traditions they use may change with them). This means that *judging a historical process one may use an as yet unspecified and unspecifiable practice.* One may base judgements and actions on standards that cannot be specified in advance but are intro-duced by the very judgements (actions) they are supposed to guide and one may even act without any standards, simply following some natural inclination. The fierce warrior who cures his wounded enemy instead of killing him has no idea why he acts as he does and gives an entirely erroneous account of his reasons. But his action introduces an age of collaboration and peaceful competition instead of permanent hostility and so creates a new tradition of commerce between nations. The question – how will you decide what path to choose? How will you know what pleases you and what you want to reject? has therefore at least two answers viz. (1) there is no decision but a natural development leading to traditions which in retrospect give reasons for the action had it been a decision in accordance with standards or (2) to ask how one will judge and choose in as yet unknown surroundings makes as much sense as to ask what measuring instruments one will use on an as yet unknown planet. Standards which are intellectual measuring instruments often have to be *invented*, to make sense of new historical situations just as measuring instruments have constantly to be invented to make sense of new physical situations.

vi. There are therefore at least *two different ways of collectively deciding an issue* which I shall call a *guided exchange* and an *open exchange* respectively.

In the first case some or all participants adopt a well specified tradition and accept only those responses that correspond to its standards. If one party has not yet become a participant of the chosen tradition he will be badgered, persuaded, 'educated' until he does – and then the exchange begins. Education is separated from decisive debates, it occurs at an early stage and guarantees that the grownups will behave properly. A *rational debate* is a special case of a guided exchange. If the participants are rationalists then all is well and the debate can start right away. If only some participants are rationalist and if they have power (an important consideration!) then they will not take their collaborators seriously until they have also become rationalists: a society based on rationality is not entirely free; one has to play the game of the intellectuals.[14]

An open exchange, on the other hand, is guided by a pragmatic philosophy. The tradition adopted by the parties is unspecified in the beginning and develops as the exchange goes along. The participants get immersed into each others' ways of thinking, feeling, perceiving to such an extent that their ideas, perceptions, world views may be entirely changed – they become different people participating in a new and different tradition. An open exchange respects the partner whether he is an individual, or an entire culture while a rational exchange promises

---

[14] 'It is perhaps hardly necessary to say' says John Stuart Mill 'that this doctrine (pluralism of ideas and institutions) is meant to apply only to human beings in the maturity of their faculties' – i.e. to fellow intellectuals and their pupils. 'On Liberty' in M. Cohen (ed.) *The Philosophy of John Stuart Mill*, New York 1961, p. 197.

respect only within the framework of a rational debate. An open exchange has no organon though it may invent one, there is no logic though new forms of logic may emerge in its course.

vii. *A free society is a society in which all traditions are given equal rights, equal access to education and other positions of power*. This is an obvious consequence of i., ii. and iii. If traditions have advantages only from the point of view of other traditions then choosing one tradition as a basis of a free society is an arbitrary act that can be justified only by resort to power. A free society thus cannot be based on any particular creed; for example, it cannot be based on rationalism or on humanitarian considerations. The basic structure of a free society is a *protective structure*, not an ideology, it functions like an iron railing not like a conviction. But how is this structure to be conceived? Is it not necessary to *debate* the matter or should the structure be simply *imposed*? And if it is necessary to debate the matter then should this debate not be kept free from subjective influences and based on 'objective' considerations only? This is how intellectuals try to convince their fellow citizens that the money paid to them is well spent and that their ideology should continue to assume the central position it now has. I have already exposed the errors-cum-deceptions behind the phrase of the 'objectivity of a rational debate': the standards of such a debate *are not* 'objective', they only *appear to be* 'objective' because reference to the group that profits from their use has been omitted. They are like the invitations of a clever tyrant who instead of saying 'I want you to do . . .' and 'I and my wife want you to do . . .' says 'What all of us want is . . .' or 'what the gods want of us is . . .' or, even better 'it is rational to do . . .' and so seems to leave out his own person entirely. It is somewhat depressing to see how many intelligent people have fallen for such a shallow trick. We remove it by observing:

viii. that *a free society will not be imposed but will emerge only where people solving particular problems in a spirit of collaboration introduce protective structures of the kind alluded to*. Citizen initiatives on a small scale, collaboration between nations on a larger scale are the developments I have in mind.

ix. *The debates settling the structure of a free society are open debates not guided debates*. This does not mean that the concrete developments described under the last thesis *already use* open debates, it means that they *could use* them and that rationalism is not a necessary ingredient of the basic structure of a free society.

The results for science are obvious. Here we have a particular tradition, 'objectively' on par with all other traditions (theses i. and vii.). Its results will appear magnificent to some traditions, execrable to others, barely worth a yawn to still further traditions. Of course, our well conditioned materialistic contemporaries are liable to burst with excitement over events such as the moonshots, the double helix, non-equilibrium thermodynamics. But let us look at the matter from a different point of view, and it becomes a ridiculous exercise in futility. It needed billions of dollars, thousands of well trained assistants, years of hard work to enable some inarticulate and rather limited contemporaries[15] to perform a few graceless hops in a place nobody in his right mind would think of visiting – a dried out, airless, hot stone. But mystics, using only their minds travelled across the celestial spheres to God himself whom they viewed in all his splendour receiving strength for continuing their lives and enlightenment for themselves and their fellow men. It is only the illiteracy of the general public and of their stern trainers, the intellectuals, and their amazing lack of imagination that makes them reject such comparisons without further ado. A free society does not object to such an attitude but it will not permit it to become a basic ideology either.

ix. *A free society insists on the separation of science and society*. More about this topic in Part Two.

3. On the Cosmological Criticism of Standards

I shall now illustrate some of these results by showing how standards are and have been criticized in physics and astronomy and how this procedure can be extended to other fields.

Section 2 started with the general problem of the relation between reason and practice. In the illustration Reason becomes scientific rationality, practice the practice of scientific research, and the problem is the relation between scientific rationality and Research. I shall discuss the answers given by idealism, naturalism and by a third position, not yet mentioned, which I shall call naive anarchism.

According to *idealism* it is rational (proper, in accordance with the will of the gods – or whatever other encouraging words are being used to befuddle the natives) to do certain things – *come what may*. It is rational (proper etc.) to kill the enemies of the faith, to avoid ad hoc hypotheses, to despise the desires of the body, to remove inconsistencies, to support progressive research programmes and so on. Rationality (justice, the

---

[15] Cf. Norman Mailer, *Of A Fire on The Moon*, London 1970.

Divine Law) are universal, independent of mood, context, historical circumstances and give rise to equally universal rules and standards.

There is a version of idealism that seems to be somewhat more sophisticated but actually is not. Rationality (the law etc.) is no longer said to be universal, but there are universally valid conditional statements asserting what is rational in what context and there are corresponding conditional rules.

Many reviewers have regarded me as an idealist in the sense just described with the proviso that I try to replace familiar rules and standards by more 'revolutionary' rules such as proliferation and counter-induction and almost everyone has ascribed to me a 'methodology' with 'anything goes' as its one 'basic principle'. But on page 32 of *AM* I say quite explicitly that 'my intention is not to replace one set of general rules by another such set: my intention is, rather, to convince the reader that *all methodologies, even the most obvious ones, have their limits*' or, to express it in terms just explained my intention is to show that idealism, whether of the simple or of the context-dependent kind, is the wrong solution for the problems of scientific rationality. These problems are not solved by a change of standards but by taking a different view of rationality altogether.

Idealism can be dogmatic and it can be critical. In the first case the rules proposed are regarded as final and unchangeable; in the second case there is the possibility of discussion and change. But the discussion does not take practices into account – it remains restricted to an abstract domain of standards, rules and logic.

The limitation of all rules and standards is recognized by *naive anarchism*. A naive anarchist says (a) that both absolute rules and context dependent rules have their limits and infers (b) that all rules and standards are worthless and should be given up. Most reviewers regard me as a naive anarchist in this sense overlooking the many passages where I show how certain procedures *aided* scientists in their research. For in my studies of Galileo, of Brownian motion, of the Presocratics I not only try to show the *failures* of familiar standards, I also try to show what not so familiar procedures did actually *succeed*. I agree with (a) but I do not agree with (b). I argue that all rules have their limits and that there is no comprehensive 'rationality', I do not argue that we should proceed without rules and standards. I also argue for a contextual account but again the contextual rules are not to *replace* the absolute rules, they are to *supplement* them. Moreover, I suggest a new *relation* between rules and practices. It is this relation and not any particular rule-content that characterizes the position I wish to defend.

This position adopts some elements of *naturalism* but it rejects the naturalist philosophy. According to naturalism rules and standards are obtained by an analysis of traditions. As we have seen the problem is which tradition to choose. Philosophers of science will of course opt for science as their basic tradition. But science is not *one* tradition, it is *many* and so it gives rise to many and partly incompatible standards (I have explained this difficulty in my discussion of Lakatos, *AM*, Chapter 16).[16] Moreover the procedure makes it impossible for the philosopher to give reasons for his choice of science over myth or Aristotle. Naturalism cannot solve the problem of scientific rationality.

As in Section 2 we can now compare the drawbacks of naturalism and idealism and arrive at a more satisfactory view. Naturalism says that reason is completely *determined by* research. Of this we retain the idea that research can change reason. Idealism says that reason completely *governs* research. Of this we retain the idea that reason can change research. Combining the two elements we arrive at the idea of *a guide who is part of the activity guided and is changed by it*. This corresponds to the interactionist view of reason and practice formulated in Section 2 and illustrated by the example of the map. Now the interactionist view assumes two different entities, a disembodied guide on the one side and a well endowed practice on the other. But the guide seems disembodied only because its 'body' i.e. the very substantial practice that underlies it is not noticed and the 'practice' seems crude and in need of a guide only because one is not aware of the complex and rather sophisticated laws it contains. Again the problem is not the interaction of a practice with something different and external, but the development of one tradition under the impact of others. A look at the way in which science treats its problems and revises its 'standards' confirms this picture.

In physics theories are used both as descriptions of facts and as standards of speculation and factual accuracy. *Measuring instruments* are constructed in accordance with laws and their readings are tested under the assumption that these laws are correct. In a similar way theories giving rise to physical principles provide standards to judge other *theories* by: theories that are relativistically invariant are better than theories that are not. Such standards are not untouchable. They can be removed. The standard of relativistic invariance, for example, may be removed when one discovers that the theory of relativity has serious

---

[16] Cf. also my complementary account in Howson, *op. cit.*

shortcomings. Such shortcomings are occasionally found by a direct examination of the theory, for example by an examination of its mathematics, or its predictive success. More likely they will be found by the development of alternatives (cf. *AM*, Chapter 3) – they will be found by research that violates the standards to be examined.

The idea that nature is infinitely rich both qualitatively and quantitatively leads to the desire to make new discoveries and thus to a principle of content increase which gives us another standard to judge theories by: theories that have excess content over what is already known are preferable to theories that have not. Again the standard is not untouchable. It is in trouble the moment we discover that we inhabit a finite world. The discovery is prepared by the development of 'Aristotelian' theories which refrain from going beyond a given set of properties – it is again prepared by research that violates the standard.

The procedure used in both cases contains a variety of elements and so there are different ways of describing it, or reacting to it.

One element and to my mind the most important one is *cosmological*. The standards we use and the rules we recommend make sense only in a world that has a certain structure. They become inapplicable, or start running idle in a domain that does not exhibit this structure. When people heard of the new discoveries of Columbus, Magellan, Diaz they realized that there were continents, climates, races not enumerated in the ancient accounts and they conjectured there might be new continents of knowledge as well, that there might be an 'America of Knowledge' just as there was a new geographical entity called 'America' and they tried to discover it by venturing beyond the limits of the received ideas. This is how the demand for content increase arose in the first place. It arose from the wish to discover more and more of a nature that seemed to be infinitely rich in extent and quality. The demand has no point in a finite world that is composed of a finite number of basic qualities.

How do we find the cosmology that supports or suspends our standards? The reply introduces the second element that enters the revision of standards viz. *theorizing*. The idea of a finite world becomes acceptable when we have theories describing such a world and when these theories turn out to be better than their infinitist rivals. The world is not directly given to us, we have to catch it through the medium of traditions which means that even the cosmological argument refers to a certain stage of competition between theories, theories of rationality included.

Now when scientists become accustomed to treating theories in a certain way, when they forget the reasons for this treatment but simply regard it as the 'essence of science' or as an 'important part of what it means to be scientific', when philosophers aid them in their forgetfulness by systematizing the familiar procedures and showing how they flow from an abstract theory of rationality then the theories needed to show the shortcomings of the underlying standards will not be introduced or, if they are introduced, will not be taken seriously. They will not be taken seriously because they clash with customary habits and systematizations thereof.

For example, a good way of examining the idea that the world is finite both qualitatively and quantitatively is to develop an Aristotelian cosmology. Such a cosmology provides means of description adapted to a finite world while the corresponding methodology replaces the demand for content increase by the demand for adequate descriptions of this kind. Assume we introduce theories that correspond to the cosmology and develop them in accordance with the new rules. What will happen? Scientists will be unhappy for the theories have unfamiliar properties. Philosophers of science will be unhappy because they introduce standards unheard of in their profession. Being fond of surrounding their unhappiness with long arias called 'reasons' they will go a little further. They will say that they are not merely unhappy, but have 'arguments' for their unhappiness. The arguments in most cases are elaborate repetitions and variations of the standards they grew up with and so their cognitive content is that of 'But the theory is ad hoc!' or 'But the theories are developed without content increase!'. And all one hears when asking the further question why that is so bad is either that science has proceeded differently for at least 200 years[17] or that content increase solves some problems of confirmation theory.[18] Yet the question was not what science does but how it can be improved and whether adopting some confirmation theories is a good way of learning about the world. No answer is forthcoming. And so some interesting possibilities of discovering the faults of popular standards are removed by firmly insisting on the status quo. It is amusing to see that such insistence becomes the more determined the more 'critical' the philosophy that is faced with the problem. We, on the other hand, retain the lesson that *the validity, usefulness, adequacy of popular standards can be tested only by research that violates them.*

A further example, to illustrate the point. The idea that information concerning the external world travels undisturbed via the senses into the

---

[17] For references and criticism cf. the article in n. 3, p. 15, as well as Chapter 16 of *AM*.
[18] John Watkins in a 'Position paper' on Critical Rationalism.

mind leads to the standard that all knowledge must be checked by observation: theories that agree with observation are preferable to theories that do not. The standard is in need of replacement the moment we discover that sensory information is distorted in many ways. We make the discovery when developing theories that conflict with observation and finding that they are excellent in many other respects (in Chapters 5 to 11 of *AM* I show how Galileo made the discovery).

Finally, the idea that things are well defined and that we do not live in a paradoxical world leads to the standard that our knowledge must be self consistent. Theories that contain contradictions cannot be part of science. This apparently quite fundamental standard which many philosophers accept as unhesitatingly as Catholics once accepted the dogma of the immaculate conception of the Virgin loses its authority the moment we find that there are facts whose only adequate description is inconsistent and that inconsistent theories may be fruitful and easy to handle while the attempt to make them conform to the demands of consistency creates useless and unwieldy monsters.[19]

The last example raises further questions which are usually formulated as objections against it (and against the criticism of other standards as well, standards of content increase included).

One objection is that non-contradiction is a necessary condition of research. A procedure not in agreement with this standard is not research – it is chaos. It is therefore not possible to examine non-contradiction in the manner described in the last example.

The main part of the objection is the second statement and it is usually supported by the remark that a contradiction implies every statement. This it does – but only in rather simple logical systems. Now it is clear that changing standards or basic theories have repercussions that must be taken care of. Admitting velocities larger than the velocity of light into relativity and leaving everything else unchanged gives us some rather puzzling results such as imaginary masses and velocities. Admitting well-defined positions and momenta into the quantum theory and leaving everything else unchanged creates havoc with the laws of interference. Admitting contradictions into a system of ideas allegedly connected by the laws of standard logic and leaving everything else unchanged makes us assert every statement. Obviously we shall have to make some further changes, for example we shall have to change some rules of derivation in the last case. Carrying out the change removes the problems and research

---

[19] For details on this point cf. Part Three, Chapter 4, Section 2, Thesis 4.

can proceed as planned.[20]

But – and with this starts another objection: how will the results of this research be evaluated if fundamental standards have been removed? For example, what standards show that research in violation of content increase leads to theories which are '*better* than their infinitist rivals' as I said a few paragraphs ago? Or what standards show that theories in conflict with observations have something to offer while their observationally impeccable rivals have not? Does not a decision to accept unusual theories and to reject familiar ones assume standards and is it not clear, therefore, that cosmological investigation cannot try to provide alternatives to all standards? These are some of the questions one hears with tiring regularity in the discussion of 'fundamental principles' such as consistency, content increase, observational adequacy, falsifiability, and so on. It is not difficult to answer them.

It is asked how research leading to the revision of standards is to be evaluated. For example, when and on what grounds shall we be satisfied that research containing inconsistencies has revealed a fatal shortcoming of the standard of non-contradiction? The question makes as little sense as the question what measuring instruments will help us to explore an as yet unspecified region of the universe. We don't know the region, we cannot say what will work in it. If we are really interested then we must either enter the region, or start making conjectures about it. We shall then find that an answer is not easy to come by and that it may need considerable ingenuity to arrive at only half way satisfactory suggestions (as an example consider the question how to measure temperature in the centre of the sun, put at about 1820); in the end somebody may come forth with an entirely unexpected solution, contrary to known natural laws, and still succeed. The very same applies to standards. Standards are intellectual measuring instruments; they give us readings not of temperature, or of weight, but of the properties of complex sections of the historical process. Are we supposed to know them even before these sections have been presented in detail? Or is it assumed that history, and especially the history of ideas is more uniform than the material part of the universe? That man is more limited than the rest of nature? Of course, education often puts limits into minds – but our problem is the *adequacy* of such limits and to examine *that* we must proceed beyond them. We therefore find ourselves in exactly the same position as the scientist with his measuring instruments – we cannot solve our problem before we

[20] In Section 3 of Chapter 4 I argue that scientific research proceeds in accordance with a practical logic whose rules of derivation don't make contradictions produce everything.

know its elements. We cannot specify standards before we know the subject matter the standards are supposed to judge. Standards are not eternal arbiters of research, morality, beauty preserved and presented by an assembly of high priests that is protected from the irrationality of the common rabble in science, the arts, in society; they are instruments provided for certain purposes by those who are familiar with the circumstances and who have examined them in detail. A scientist, an artist, a citizen is not like a child who needs papa methodology and mama rationality to give him security and direction, he can take care of himself, for he is the inventor not only of laws, theories, pictures, plays, forms of music, ways of dealing with his fellow man, institutions *but also of entire world views, he is the inventor of entire forms of life*. The questions only reveal the disorientation of those not familiar with the structure and the problems of concrete research.[21] For them research is like a game for children that proceeds in accordance with a few simple rules known to the parents who therefore can kindly but firmly point out whenever there has been a violation of rules. Philosophers of science like to view themselves as such parents. Small wonder they get confused when their authority is challenged.

The habit, started by the Vienna Circle and continued by critical rationalists to 'translate' problems into the 'formal mode of speech' has greatly contributed to the protection of basic standards of rationality. Take again the question of the finiteness vs. infinitude of the world. This question, so it seems, is a factual question, to be solved by research. To 'make it clearer and more precise' (a famous phrase used by positivists and critical rationalists when replacing complex problems they do not understand by simplistic caricatures they can make sense of) it is 'translated' into a property of sequences of explanations. In the one case (finite universe) there is one 'basic' explanation or 'ultimate' explanation on which all other explanations depend. In the other case (infinite universe) we have not a single explanation but an infinite and never ending sequence. Critical rationalists have given abstract reasons why such sequences are to be preferred. They are to be preferred they say, because they conform to the 'critical attitude' recommended by that school. Now if the cosmological background is forgotten, then this already decides the matter: there are no basic explanations. Popper goes even further. Declaring 'that the world of each of our theories may be explained, in

---

[21] Cf. Thesis 5 of the preceding section and the corresponding explanations.

turn, by further worlds which are described by further theories'[22] he concludes that 'the doctrine of an essential or ultimate reality collapses'. It collapses – why? Because it is inconsistent with Popper's favourite methodology. But if the world is finite then there exists an ultimate reality and critical rationalism is the wrong philosophy for it.

The issue between realism and instrumentalism gives rise to similar observations. Do electrons exist or are they merely fictitious ideas for the ordering of observations (sense data, classical events)? It would seem that the question has to be decided by research (cf. also the remarks in Section 3 of Part Three, Chapter Four, below). *Research* has to decide whether there are only sensations in this world, or whether the world contains also more complex entities such as atoms, electrons, living beings and so on. If there are only sensations, then terms such as 'electron' or 'St. Augustine' are auxiliary terms, designed to bring some order into our experiences. They are like operators in mathematics, or connectives in logic, they connect assertions about sense data, they do not refer to things different from sense data. Modern professional realists do not see matters in this way. For them the interpretation of theories can be decided on purely methodologically and independently of scientific research. Small wonder their notion of reality and that of the scientists have hardly anything in common.[23]

4. 'Anything Goes'

One way of criticizing standards is to do research that violates them (this is explained in Section 3). In evaluating the research we may participate in an as yet unspecified and unspecifiable practice (this is explained in Section 2, Thesis v.). Result: interesting research in the sciences (and, for that matter, in any field) often leads to an unpredictable revision of standards though this may not be the intention. *Basing our judgement on accepted standards* the only thing we can say about such research is therefore: anything goes.

Note the context of the statement. 'Anything goes' is *not* the one and only 'principle' of a new methodology, recommended by me. It is the only way in which those firmly committed to universal standards and

---

[22] 'Three Views Concerning Human Knowledge' quoted from *Conjectures and Refutations*, London 1963, p. 115.

[23] For details cf. Chapter 5 of my *Der Wissenschaftstheoretische Realismus und die Autorität der Wissenschaften*, Wiesbaden 1978.

wishing to understand history in their terms can describe my account of traditions and research practices as given in Sections 2 and 3. If this account is correct then all a *rationalist* can say about science (and about any other interesting activity) is: anything goes.

It is not denied that there exist parts of science that have adopted some rules and never violate them. After all, a tradition can be streamlined by determined brainwashing procedures and it will contain stable principles once it has been streamlined. My point is that streamlined traditions are not too frequent and that they disappear at times of revolution. I also assert that streamlined traditions accept standards without examining them and that any attempt to examine them will at once introduce the 'anything goes' situation (this was explained in Section 3).

Nor is it denied that the proponents of change may have excellent arguments for every one of their moves.[24] But their arguments will be *dialectical arguments*, they will involve a changing rationality and not a fixed set of standards and they are often the first steps towards introducing such a rationality. This, incidentally, is also the way in which intelligent commonsense reasoning proceeds – it may start from some rules and meanings and end up with something entirely different. Small wonder that most revolutionaries have unusual developments and often regard themselves as dilettantes.[25] It is surprising to see that philosophers who were once inventors of new world views and who taught us how to look through the status quo have now become its most obedient servants: *philosophia ancilla scientiae*.

5. The 'Copernican Revolution'

In *AM* I used Galileo as an illustration of the abstract principles just explained. But the 'Copernican Revolution' does not only contain Galileo. It is a very complex phenomenon. To understand it one must divide the knowledge of the time into different and often fairly independent components, one must examine how different groups at different times reacted to each component and so slowly build up the process which today is called, rather summarily, the 'Copernican Revolution'. Only such a piece-by-piece study will provide us with information about reason and practice that is not merely a repetition of our methodological

---

[24] Cf. Section 9 of 'Consolations for the Specialist' in Lakatos and Musgrave *Criticism and the Growth of Knowledge*, Cambridge 1970.

[25] Bohr, Einstein, Born regarded themselves as dilettantes and often said so.

dreams.

It is also necessary to state clearly what one wants to know.

I have chosen the following three questions which seem to be of general interest.

A. Are there rules and standards which are 'rational' in the sense that they agree with some plausible general principles and demand attention under all circumstances, which are obeyed by all good scientists when doing good research and whose adoption explains events such as 'the Copernican Revolution'?

The question is not merely whether a sequence of events such as: suggestion of theory T – occurrence of certain phenomena – acceptance of the theory agrees with some standards, it asks in addition whether the standards were consciously used by the participants. We hardly shall call people rational who act rationally in *our* sense but achieve this by bungling standards *they* regard as important. Neglect of this point is a decisive drawback of the otherwise excellent paper by Lakatos and Zahar.[26]

B. Was it reasonable, at a given time, to accept the Copernican point of view and what were the reasons? Did the reasons vary from one group to the next? From one period to the next?

C. Was there a time when it became unreasonable to reject Copernicus? Or is there always a point of view that permits us to regard the idea of a motionless earth as a reasonable idea?

It seems that the answer to A is no, the answer to B yes (for all questions) and the answer to C a qualified yes (for both questions). I shall now give a sketch of the arguments that lead to such a result.

First, the general talk of a 'revolution in astronomy' must be replaced by an analysis of elements that can be identified. We must distinguish:

1. The situation in cosmology
2.   physics
3.   astronomy
4.   tables
5.   optics
6.   theology

The distinctions are not made 'to be precise' – they reflect the actual historical situation. For example, 1 depended on 2, but not entirely so. This becomes clear in the 17th century. 3 was independent of 1 and 2 as well as of 5, 4 depended on 3 but some additional information was

---

[26] 'Why did Copernicus supersede Ptolemy?' in R. S. Westman (ed.) *The Copernican Achievement*, University of California Press 1974.

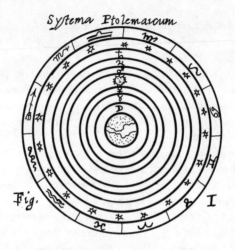

needed. Finally, 6 provided a boundary condition for 1 and 2, but not for 3.

The situation is reflected in the textbooks. Sacrobosco and his imitators give a sketch of 1, hardly mention 2, give only an account of the main circles of the sky for 3 and omit 4, 5 and 6. Handbooks on astronomy such as Ptolemy's own magnificent work contain 3 and 4 but only the bare elements of 1 and 2 are mentioned, and then in a most perfunctory manner. The same is true of 5. Textbooks of physics discuss 2, elements of 1 but not 3, 4, 5 or 6. Philosophers explain that the task of 2 is to give a true description of the processes of this world and of the laws that guide them while the task of 3 is to provide correct predictions, by whatever means possible. An astronomer, it is said, is not concerned with truth, he is concerned with predictions.[27] All he can claim for the ideas he uses is that they produce such predictions; he cannot claim that they are true. There were many thinkers, mainly among the Arabs, who tried to give physical explanations of the success of certain astronomical devices. To some extent we may compare them with those who tried to explain the laws of phenomenological thermodynamics with the help of the atomic theory.

The basic assumption of 1 was the central symmetrical universe – the earth in the centre surrounded by a variety of spheres up to and including the sphere of the fixed stars. The earth is at rest, it neither rotates nor

---

[27] For a more detailed account and many quotations cf. P. Duhem, *To Save the Phenomena*, University of Chicago Press 1972.

moves in any other way. There are two kinds of basic motions in this universe, sublunar motions, i.e. motions of things below the moon, and

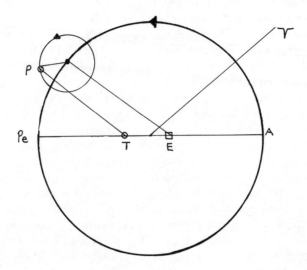

superlunar motions, i.e. motions of things above the moon. *Undisturbed* sublunar motions depend on the element moving: fire and air move up, water and earth down, though with different intensities. The motion of a 'mixed' body depends on the percentage of the elements in it.[28] All superlunar motions are circular. Arguments for these assertions are found in Aristotle's own *On the Heavens* and are repeated without much refinement in later textbooks.[29]

The basic assumptions of 2 are that every object consists of matter and form, that change involves an exchange of forms, that it is due to outer influences (if there is no outer influence then everything remains unchanged) and proportional to the strength of this influence (and the inverse of the resistance). The assumptions are argued for in Aristotle's *Physics* and are again repeated without much refinement in later textbooks.[30]

[28] According to this interesting theory a body is defined not by its *substance* but by its *motion*. Modern elementary particle physics has resumed this approach.

[29] Cf. T. Kuhn, *The Copernican Revolution*, Cambridge 1967.

[30] But there are many discussions in commentaries. Cf. Marshall Clagett *The Science of Mechanics in the Middle Ages*, Madison 1964.

The theory of motion of Aristotelian physics comprises not only locomotion but all kinds of change. It was used and still is used in subjects such as biology, medicine, physiology, bacteriology to discover 'disturbing entities' such as eggs of flies, bacteria, viruses and so on. Newton's law of inertia is of no help in these domains.

Again the basic assumptions are supported by arguments which are either empirical, or logical or both. The main purpose of the arguments is to show that the commonsense view of the world as expressed in our perception and codified in our language is basically correct though there are disturbances which can be studied and removed. The commonsense view is not simply *accepted*, there are *arguments* to show why it can be trusted. Details will be explained in the next section.

The basic assumptions of 3 are shown in the above model. Venus, Mars, Jupiter and Saturn are assumed to move on a small circle, the so-called epicycle whose centre moves on a larger circle, the so-called deferent. The motion on the deferent has constant angular velocity not around its centre, but around the point E, the *equant*. The planet is seen from the earth T, equally distant from the centre as E, but on the other side of it. It moves on its epicycle with constant angular velocity in such a manner that the radius vector from the centre of the epicycle to the planet is parallel to the mean longitude of the sun. There is one such scheme with different constants for each of the four planets mentioned. The sun, the moon, and Mercury are treated in a different way. Planetary latitude is determined independently according to a scheme I shall not mention here.

It has been calculated [31] that *given the proper constants* this scheme 'can account for all angular motions of the planets to an accuracy better than 6' ... excepting only the special theory needed to account for ... Mercury and excepting also the Planet Mars which shows deviations up to 30' from such a theory. [This is] certainly better than the accuracy of 10' which Copernicus himself stated as a satisfactory goal for his own theory' which was difficult to test especially in view of the fact that refraction (almost 1° at the horizon) was not taken into account at the time of Copernicus and that the observational basis of the predictions was less than satisfactory.

To calculate 4 one needs additional constants such as the latitude of the place from which the observations are made. Thus 4 may contain

[31] Derek J. de S. Price 'Contra Copernicus' in *Critical Problems of the History of Science*, ed. Clagett, Madison 1959, pp. 197–218. For the versatility of the Ptolemaic scheme cf. also Hanson, *Isis*, No. 51, 1960, pp. 150–8.

errors that cannot be projected back upon the basic theory. Ptolemaean predictions were often way off the mark because of an incorrect choice of constants. It is therefore not reasonable to remove 3 because of gross conflicts with observation.

5 entered astronomy only with the telescope. The story has been told in *AM*. Additional details will be given in the next section.

6 is hardly ever mentioned by modern philosophers though it played a decisive role in the debate. The attitude of the Church was not as dogmatic as is often assumed. Interpretations of Bible passages had been revised in the light of scientific research before. Everyone regarded the earth as spherical and as freely floating in space though the Bible tells a very different story. But the arguments of the Copernicans, those of Galileo included simply *were not regarded* as decisive. Nor *were* they that decisive as was shown in *AM*. The Bible still played an important role for Newton who used both God's work and God's word to explore the plans of God.[32] In the sixteenth century agreement with the Word of God as contained in Holy Scripture was an important and universally accepted boundary condition of physical research. It was a standard comparable to the 'modern' standard of experimental accuracy.

There were three arguments against the motion of the earth. The first argument, the so-called tower argument (and other arguments of the same kind), came from physics. It is explained and discussed in *AM*, pp. 70ff. The argument rests on the Aristotelian theory of motion which was confirmed by experience.

The second argument, already mentioned by Aristotle, is the argument from parallax: if the earth moves around the sun then traces of this motion must be found in the stars. No such traces were found.

The third argument was that the motion of the earth was in conflict with the Bible. In the debate about Copernicus all these arguments were used, but the first and the third were regarded as more important than the second.

Today we have the following theories of the transition from Ptolemy/ Aristotle to Copernicus/Galileo.

1. Naive empiricism: in the 'Middle Ages' one paid attention to the Bible, but then people raised their heads, watched the sky and discovered that the world was different from what they had thought it was. This theory has now almost disappeared. One occasionally finds

---

[32] Cf. F. Manuel *The Religion of Isaac Newton*, Oxford 1974, where further literature is given. Cf. also the chapter on Newton in A. Koyré, *From the Closed World to the Infinite Universe*, Cambridge 1964.

it, as an aside, in books on the history of literature.

2. Sophisticated empiricism: new observations were made that forced astronomers to revise an already empirical astronomy.

3. Conventionalism: the old astronomy became more and more complicated and so it was finally replaced by a simpler account.

4. Falsificationism: new observations refuted some decisive assumptions of the old astronomy, and so a new astronomy had to be found.

5. Crisis Theory: astronomy was in a crisis and the crisis had to be dealt with. This is Kuhn's theory.

6. Research programme account: the Ptolemaic Research Programme degenerated while the Copernican research programme progressed.

All these theories have certain assumptions in common. They can be criticized already because of these assumptions, for the assumptions are quite implausible.

For example, it is assumed that a complex process involving experts from different and partly independent fields with different and partly independent standards can be explained by adoption of a single standard. It is also assumed that this standard was accepted before, during and after the conflagration, that it was the principle that turned the participants against the status quo and guided them in their search for something better. The last assumption certainly is not correct. Ptolemean astronomers regarded degeneration not as an objection, but as a sign of excellence: the ancient principle that astronomy must 'save the phenomena' means that it must 'degenerate' in the sense of Lakatos. Hence, if Copernicanism was accepted because it was 'progressive' then the acceptance involved a change of theories *as well as of* standards and was therefore *not* 'rational' in the sense of Lakatos (and of theory 6). Thirdly, most of the accounts consider only astronomy and disregard the other subjects that participated in the change and were changed as a result. We see: it does not need detailed research to suspect that the proposed theories cannot possibly be true. A closer look confirms this suspicion.

1, 2, 4 and 5 assume that new observations were made in the first third of the 16th century, that these observations showed the inadequacy of the Ptolemaic scheme, that the inadequacies were removed by Copernicus which was the reason why Copernicus superseded Ptolemy. The assumptions are meant to apply to astronomy only and so only astronomy will be discussed. Is it true that there were new observations in this science, that the observations presented problems and that Copernicus solved the problems?

One way of answering the question is to look at the *tables*. Were the post-Copernican tables better than their predecessors? Gingerich who has examined the matter[33] says they were not: mean errors and maximal errors are about the same, but they are distributed in different ways and show a different pattern. This was already noticed in the 16th century: the *Prutenic tables* were not much better than the *Alfonsine tables.*

Another way of answering the question is to consult the participants. Now Copernicus, far from criticizing Ptolemy for a failure to make correct predictions calls his theory 'consistent with the numerical data'.[34] And instead of enumerating the new observations that had prompted him to revise astronomy he says that 'we must follow in their [the ancient Greeks'] footsteps and hold fast to their observations bequeathed to us like an inheritance. And if anyone on the contrary thinks that the ancients are untrustworthy in this regard, surely the gates of this art are closed to him.'[35] Neither new observations nor the inability of Ptolemy to take care of the old observations are the reason for Copernicus' research. This disposes of 1, 2, 4 and 5 at least as far as Copernicus himself is concerned.

Naive empiricism has further drawbacks. It overlooks that Aristotle is an arch empiricist and it also overlooks how carefully Copernicus, Tycho, Galileo and others discuss the theological arguments against the motion of the earth.

Conventionalism fails because Copernicus' final system is hardly less complicated (in terms of numbers of epicycles) than that of Ptolemy. A look at the drawings of the two systems makes that very clear.[36]

The research programme account fails because astronomers and physicists did not consider and accept Copernicus for the reasons given by this theory. Also acceptance should have started immediately after Copernicus' main opus became known – but it did not. Nobody was then 'rational' in the sense of Lakatos and Zahar.

3, 4 and 6 also omit the difficulties created by physics and theology. Today only a few would accept a theory in conflict with energy conservation just because it was simple. Why should astronomers in the 16th century have accepted a physically and theologically impossible theory just because of its simplicity? Similar questions can be raised against

---

[33] 'Crisis vs. Aesthetics in the Copernican Revolution', *Vistas in Astronomy*, Vol. 17, ed. Beer, 1974. Gingerich compares the tables of Stoeffler with those of Stadius, Maestlin, Magini and Origanus.

[34] *Commentariolus*, quoted from E. Rosen (ed.), *Three Copernican Treatises*, New York 1959, p. 57.

[35] *Letter Against Werner* in Rosen, *op. cit.*, p. 99.

[36] For the drawings cf. de Santillana's edition of Galileo's *Dialogue*, Chicago 1964.

4 and 6. With 4 there is the additional comment that Copernicus was refuted by facts such as the behaviour of falling stones while Ptolemy/ Aristotle was not. We see: the theories so far introduced to explain the Copernican Revolution are implausible in their general assumptions and false in their details. They are based on wrong views concerning the relation between reason and practice.

That there must be something wrong with the belief that the Copernican view had advantages over its rivals and that these advantages were noted at the time becomes clear when reading the following passage from Galileo's *Dialogue Concerning the Two Chief World Systems*. In this *Dialogue* Salviati who 'act[s] the part of Copernicus'[37] replies to Sagredo who had expressed his astonishment at the small number of Copernicans. 'You wonder' he says 'that there are so few followers of the Pythagorean opinion [that the earth moves] while I am astonished that there have been any up to this day who have embraced and followed it. Nor can I ever sufficiently admire the outstanding acumen of those who have taken hold of this opinion and accepted it as true: they have, through sheer force of intellect done such violence to their own senses as to prefer what reason told them over that which sensible experience plainly showed them to be the contrary. For the arguments against the whirling [the rotation] of the earth . . . are very plausible as we have seen; and the fact that the Ptolemaics and the Aristotelians and all their disciples, took them to be conclusive is indeed a strong argument of their effectiveness. But the experiences which overtly contradict the annual movement [the movement of the earth around the sun] are indeed so much greater in their apparent force that, I repeat, there is no limit to my astonishment when I reflect that Aristarchus and Copernicus were able to make reason so conquer sense that, in defence of the latter, the former became mistress of their belief.'[38]

A little later Galileo notes that 'they [the Copernicans] were confident of what their reason told them.'[39] And he concludes his brief account of the origin of Copernicanism by saying that 'with reason as his guide he [Copernicus] resolutely continued to affirm what sensible experience seemed to contradict.' 'I cannot get over my amazement' Galileo-Salviati repeats 'that he was constantly willing to persist in saying that Venus might go around the sun and might be more than six times as far from us at another, and still look equal, when it should have appeared

[37] *Dialogue*, tr. Stillman Drake, University of California Press 1953, pp. 131 and 256.
[38] p. 328.
[39] p. 335.

forty times larger.'[40]

This is how the matter looked even at the beginning of the 17th century.[41] It is clear that most of the simple philosophical theories mentioned above must be replaced by different and more realistic accounts.

To arrive at such accounts, I shall proceed in small steps, consulting only Copernicus' own writings and those contemporaries who were familiar with them.

First, Copernicus.[42] It seems that the basic motive of Copernicus was the restoration of Greek astronomy. 'The planetary theories of Ptolemaics and most other astronomers . . . seemed . . . to present no small difficulty. For these theories were not adequate unless certain equants were also conceived; it then appeared that a planet moved with uniform velocity neither along its own deferent nor relative to an actual centre. . . . Having become aware of these defects I often considered whether there could perhaps be found a more reasonable arrangement of circles from which every apparent inequality could be derived and in which everything would move uniformly about its proper centre as the rule of accomplished motion requires. . . .'

In this quotation we have a distinction between apparent motion and real motion and the task of astronomy is conceived as explaining the former ('every apparent inequality') in terms of the latter. Ptolemy, Copernicus says, does not carry out the task, for he uses equants. Equants predict apparent motion (the inequalities of the planet along its deferent) not in terms of real motion but in terms of other apparent motions where 'a planet move[s] with uniform velocity neither along its own deferent nor relative to the actual centre'. Real celestial motion for Copernicus as well as for the Ancients is uniform circular motion about a centre. It is in terms of such motion that the inequalities must be explained.

Copernicus removes excentres and equants and replaces them by two epicycles for each planet. Having thus already populated the deferent he must try to explain the synodic anomaly (the stations and retrogressions)

---

[40] p. 339. Galileo here refers to the fact that Venus, because of the varying distance from the earth should have varied much more in its brightness than it actually did. Cf. Appendix i of *AM* for that point. According to Galileo there existed therefore two kinds of arguments against the motion of the earth: dynamical arguments, taken from Aristotle's theory of motion; and optical arguments. He tried to remove them both.

[41] But we must not overlook Galileo's rhetoric which made the difficulty appear more pressing in order to make his solution appear more ingenious.

[42] In what follows I accept the account given by Fritz Krafft 'Copernicus Retroversus i and ii', *Colloquia Copernicana* iii and iv, *Proceedings of the Joint Symposium of the IAU and IUHPS*, Torun 1973. Translation of the passages (from the *Commentariolus*) by Rosen *op. cit.*, p. 57, corrected by Krafft, i, p. 119.

in a different way. In trying to find a new explanation Copernicus made use of the fact that the synodic anomaly always agrees with the position of the sun.[43] One could therefore try to explain it as an appearance, created by a motion of the earth.

Such an explanation no longer permits us to calculate each planetary path separately, and independently of the rest for it ties all planets to the Great Circle (the path of the earth around the centre[44]) and, therefore, to each other. We have now a *system* of the planets and with it a 'design of the universe and the definite symmetry of its parts'. 'For all these phenomena' Copernicus writes in his later work[45] 'appear to be linked most nobly together, as by a golden chain; and each of the planets, by its position and order, and every inequality of its motion bears witness that the earth moves and that we who dwell upon the globe of the earth, instead of accepting its changes of position, believe that the planets wander in all sorts of motions of their own.' It is this internal connectedness of all the parts of the system together with his belief in the basic nature of circular motion that makes Copernicus pronounce the motion of the earth as real.

The motion of the earth is in conflict with cosmology, physics and theology (in the sense in which these subjects were understood at the time – see above). Copernicus removes the conflict with theology by a familiar device: the word of scripture is not always to be understood literally. He resolves the conflict with physics by proposing his own theory of motion which agrees with some parts of the Aristotelian doctrine, but not with others.[46] The argument is surrounded by references to ancient beliefs such as Hermeticism and the idea of the exceptional role of the sun.[47]

[43] Of the *mean sun* in Copernicus. Only Kepler achieved reduction to the real sun.

[44] The centre of the world does not coincide with the sun.

[45] *De Revolutionibus* Address to Pope Paul. Fritz Krafft (n. 39) suggests that Copernicus discovered the harmony only in the course of his attempt to carry out the programme of centred circular motion. Centring of Circles was his first intention. The synodic anomaly then became a problem. It was solved by assuming a motion of the earth. This assumption tied the planetary paths together into a system and so gave rise to the 'harmony' which became a second argument, and soon the most important one.

[46] Copernicus connects the motion of the earth with its shape: the earth is spherical, hence it can (must) rotate and move in a circle. This does not take care of the two further motions C ascribed to the earth and which he needs for precession (cum trepidation) and the parallelism of the axis of the earth. Nor does it take care of the assumption, fundamental for Copernican physics that parts of the earth participate in its motion even when separated from it. The last assumption is a direct application of Aristotle's principles of celestial motion to the earth and thus obliterates the distinction between sublunar and superlunar elements and motions.

[47] Cf. *AM*, p. 95, n. 12.

The argument is convincing only to those who prefer mathematical harmony to an agreement with the qualitative aspects of nature or, to express it differently, who incline towards a Platonic rather than an Aristotelian interpretation of nature. The preference is 'objective' only if there are 'objective' reasons for Platonism and against Aristotelianism.[48] But one knows quite well that harmony can be a harmony of appearances (cf. Plato on the lawful foreshortenings that were compensated by 'false' proportions of statues and pillars) and we have learned, especially from the quantum theory that harmonious mathematical relations such as are found e.g. in Schrödinger's theory of microparticles[49] need not reflect an equally harmonious arrangement of nature. This is what the Aristotelians asserted: the extent to which a theory reflects nature cannot be read off its structure but must be provided by a different theory that directly describes nature and Aristotle's physics is such a theory. On the other hand, there were numerous difficulties in Aristotle. Some of them concerned special phenomena such as the motion of things thrown and were not regarded as objections. Others seemed to discredit the Aristotelian system as a whole. In raising such comprehensive objections one used interpretations of Aristotle which had little to do with the author himself, which tied all his assertions, theories, arguments into a system which was then weakened by every single difficulty. The weight given to harmony, or to 'Aristotle', depended, therefore, on the attitude taken towards the difficulties and this attitude in turn depended on expectations one had concerning their removal. And as these expectations varied from one group to the next the entire argument was firmly embedded in a background that can only be called 'subjective'.[50]

Thus Copernicus, Rheticus, Maestlin took the argument from harmony as basic and so did Kepler. Tycho mentioned it, seemed to like it but did not accept it. For him the physical and the theological difficulties decided the matter.[51] The members of the Wittenberg school who studied

---

[48] I use this abbreviated manner of speaking without implying that the parties in the debate took a Platonic or an Aristotelian position in the sense of these authors and in the full knowledge of their intellectual background.

[49] I am speaking now of Schrödinger's original theory and not of the form it took when it was incorporated into the Copenhagen interpretation.

[50] One might try to 'objectivise' the expectations by reference to some 'logic of induction'. That does not do justice to the debate as the parties had also different ways of evaluating their guesses.

[51] 'Tychonis Brahei de Disciplinis mathematicis oratio publice recitata in Academia Haffniensi anno 1574 (= *Opera Omnia* Vol. i, pp. 143–73).

Copernicus in some detail were unimpressed.[52] Many of them used the Copernican arrangement and the Copernican constants as starting points but the final results were reduced to the motionless earth. Everyone praised the restoration of circularity.

Maestlin is an excellent example of an astronomer who concentrates on mathematical relations with hardly any interest in the 'physics' of his time. Astronomers do not need to examine Aristotle because they can settle problems in their own way: 'Copernicus wrote his entire book not as a physicist but as an astronomer'.[53] Mathematical reasoning is not only exact, it has its own criteria of reality: 'This argument [from harmony] is wholly in accord with reason. Such is the arrangement of this entire, immense machine that it permits surer demonstrations: indeed, the entire universe revolves in such a way that nothing can be transposed without confusion of its [parts] and hence, by means of this all, the phenomena of motion can be demonstrated most exactly, for nothing unfitting occurs in the course of their orbits'.[54] Maestlin was further strengthened in his belief when he discovered that the comet of 1577 moved in the Copernican orbit of Venus – excellent proof for the reality of these orbits.[55]

Maestlin's attitude towards Aristotle is shared by many thinkers among them artisans, scholars with wider interests, laymen with friends among artisans and scholars. Being familiar with the amazing discoveries of the century and with the difficulties these discoveries created for the received body of knowledge they put greater emphasis on transgressing boundaries than on the orderly arrangement of information inside them. The discovery of America made them suspect the existence of an America of knowledge as well and they interpreted each difficulty as evidence for this new continent rather than as a 'puzzle' to be solved by accepted methods. Problems were not treated one by one as was the habit of the Aristotelians[56] but as parts of a pattern and one extended them beyond the area of their impact to domains apparently not touched by them. This

[52] Westman, 'The Wittenberg Interpretation of the Copernican Theory', *Isis*, Vol. 33, 1972.

[53] Maestlin's marginal notes to *de Rev.* quoted after Westman, 'Michael Maestlin's adoption of the Copernican Theory', *Colloquia Copernicana* iv, Ossilineum 1975, p. 59.

[54] *loc. cit.*

[55] Cf. Westman in *Colloquia Copernicana* i, Warsaw 1972, pp. 7–30 for details. Kepler accepted the argument which made him a Copernican.

[56] This is how the Copernican view is treated in Riccioli's *Almagestum Novum*. Each single difficulty for Ptolemy/Aristotle is discussed by itself and 'solved', each single argument for the Copernican view is examined by itself and refuted. Kepler, however (letter to Herwarth quoted from Caspar-Dyck *Johannes Kepler in seinen Briefen*, Vol. i, Munich 1930, p. 68),

is how Brahe's location of the nova of 1572[57] and his discovery that comets move across the celestial spheres obtained an importance they would not have otherwise had.[58] For some Aristotle was a hindrance not only to knowledge but also to religion[59] and so they became interested in alternatives. It was this interplay of attitudes, discoveries, difficulties that gave Copernicus a more than astronomical importance and that later on removed Aristotle even from domains that not only had evidence for his views but were in need of his philosophy: his removal from astronomy sufficed to regard him as superseded. Can this judgement be accepted today? I do not think it can.

6. Aristotle not a Dead Dog

Aristotle's philosophy is an attempt to create a form of knowledge that reflects man's position in the world and guides him in his enterprises. To carry out the task he uses the achievements of his predecessors. He studies them in detail and creates a new subject – the history of ideas. He also uses commonsense which he regards as a trustworthy source of information and which he often prefers to the speculations of the intellectuals. Xenophanes, Parmenides, Melissos had discovered that concepts can be connected in special ways and had constructed new kinds of stories (today we call them 'arguments') about the nature of things. Concepts

---

emphasizes that though 'each of these reasons for Copernicus taken for itself would find only scant belief' their joint effect amounts to a strong argument. Cf. also his *Conversations with Galileo's Sidereal Messenger*, tr. E. Rosen, New York 1965, p. 14, where Kepler speaks of 'mutually self-supporting evidence'. *The transition from local arguments to arguments considering a 'consilience of inductions' (or guesses) as the matter was called much later is an important element of the 'Copernican Revolution'.* Without it the development would have been much more staid and it might not even have taken the same direction.

[57] He located it in the eighth sphere, among the fixed stars.

[58] Many contemporaries regarded the comet of 1577 as being of supernatural origin and therefore not as an objection to Aristotelian doctrine. Cf. Doris Hellman, *The Comet of 1577*, New York 1944, pp. 132, 152 and 172. Not everybody was affected by the discoveries in the same way and the arguments we hear today were not the arguments that were effective then. At any rate – they needed the background described in the text above to make an impression.

[59] The conflict between Aristotle and the Church started much earlier, when the Aristotelian writings were gradually made available in Latin. Cf. E. Grant, *A Source Book in Mediaeval Science*, Cambridge, Mass. 1974, pp. 42ff. As opposed to the theological difficulties of Copernicus the conflict was not about literal vs. non literal interpretations of *scriptiral passages* but about *basic principles*. Thus for Aristotle the world is eternal while it is created for the Church. Aristotle assumes basic principles of physics and of reasoning while the Church assumed that God could supersede any principle he wished. And so on.

entered these stories only if they furthered their swift completion which means that the stories were no longer about the familiar entities of tradition and experience. They were about 'theoretical entities'. Theoretical entities were introduced not because one had found that they existed while their traditional ancestors did not; they were introduced because they fitted into the stories while their traditional ancestors did not. Not tradition, not experience, but the 'fitting' decided about their existence. Arguing became quite popular, expecially because it soon led to absurd consequences[60] and with this the new concepts also gained in popularity. It was this popularity and not any thorough examination of the matter that decided about their fate.

Here are some examples of the arguments used.[61]

God, it is said, must be *one*. If he were many, then the many would be equal or unequal. If they are equal, then they are again one. If not, then some are, and those are one (first part) while others are not, and those do not count. Or: God cannot have *come into existence*. Had he done so then he would have emerged from what is equal, or what is not equal. Emerging from what is equal means remaining the same; emerging from what is unequal is impossible for what is cannot come from what is not. Also, God must be *all powerful*: an all powerful god comes from what is equal or from what is not equal. In the first case he again does not emerge but remains the same. In the second case he comes from what is stronger or from what is weaker. He cannot come from what is stronger, for in that case the stronger would still exist. He cannot come from what is weaker, for where should the weak obtain the strength to create the stronger?

Two elements characterize these arguments. First, the *form* which is: if A then either B or C. Neither B nor C, therefore not A. This form plays a role both in the 'sciences' (Zeno!) and in the 'arts' (in the *Oresteia* of Aeschylos Orestes runs into an impossibility whether he kills his mother or whether he does not kill her; the conundrum is thrown back on the structure of the society and solved by introducing an assembly that decides the matter[62]). The second element is the 'conservation principles' used in trying to establish not-B and not-C. According to one of

[60] For the popularity of the arguments cf. Gershenson and Greenberg, 'The "Physics" of the Eleatic School: a Reevaluation' in *The Natural Philosopher* 3, New York 1964.

[61] The arguments occur later among the Sophists, and with sceptical purpose. They are summarized in *On Melissos, Xenophanes and Giorgias* which can hardly have arisen before the 1st century. Reinhardt regards it as an essentially correct account of the dialectics of Melissos and therefore, also of Xenophanes. His reasons are convincing, but not generally accepted. Even so the book gives us some insight into popular forms of thought.

[62] Cf. the excellent analysis in von Fritz, *Antike und Moderne Tragödie*, Berlin 1962.

these conservation principles the only property a God possesses (and which can make him different from other gods) is his *being* or his *strength*. Difference means difference in being, i.e. not-being. This is a very diluted and quite inhuman conception of a deity indeed[63] at variance both with tradition and with the experience and expectations of the time. Xenophanes ridicules the traditional concepts as being anthropomorphic ('If cows had hands, they would paint the gods in their image . . .') and thereby supports the monotheistic tendencies which were quite strong at the time. But the One God of the Philosophers who gradually emerges has features that are not determined by his relation to man and the universe but by the way in which his *concept* fits into some simple types of abstract reasoning. The playful demands of the intellect come to the fore and determine what can and what cannot exist. God and Being become abstractions because these abstractions can be more easily handled by the intellect, and because it is easier to draw surprising conclusions from them. It would be of great interest to follow this development in detail and to discover how it was that a new way of juggling words could become a danger to experience and tradition.

Aristotle accepts the achievements of his predecessors, some of their methods of proof included. But he does not simplify concepts as they do. He increases their complexity to get nearer to commonsense while at the same time developing a theory of objects, change and motion that can deal with the more complex concepts. The new and philosophical commonsense that arises in this way is supported not only by the *practical authority* of the familiar commonsense that guides every step of our lives but also by the *theoretical authority* of Aristotle's considerations.[64] Commonsense is in and with us, it is the practical basis of our thoughts and actions, we live by it, but we can now demonstrate its inherent rationality as well and so have two arguments for it instead of one. Aristotle soon adds a third: extending his theory of motion to the interaction between man and the world he finds that man perceives the world as it is and so shows how intimately theoretical considerations and practical actions (processes, perceptions) are connected with each other. This delicate and complex fitting job which confirms the original belief in the harmony of man and nature is the background of Aristotle's more con-

---

[63] F. Schachermayr, *Die frühe Klassik der Griechen*, Stuttgart 1966, p. 45, speaks of Xenophanes' 'more sublime conception of God' which shows quite clearly what an intellectual means by 'sublime'.

[64] Commonsense and reason go different ways in Hume. They do not in Aristotle.

crete ideas of knowledge and being.[65] Briefly, the ideas are as follows.

According to Aristotle universals *arise* from sense experiences and principles are *tested* by comparing them with observations. This is a *physical theory* – it describes the physical process that moulds the mind and establishes universals in it. The process depends on particulars as well as on 'low level universals' already imprinted.[66] An idiosyncratic history of perception will therefore lead to idiosyncratic perceptions later on. We use the experience so constituted to 'find the principles which belong to each subject. In astronomy, for example, it was astronomical experience that provided the principles of the science, for it was only when the phenomena were adequately grasped that the proofs in astronomy were discovered. And the same is true of any science whatsoever.'[67] Accordingly, the 'loss of any one of the senses entails the loss of the corresponding portion of knowledge'.[68] Principles not consistent with observation 'are wrongly assumed . . . [for] . . . principles require to be judged by their results, and particularly from their final issue. And in the case of knowledge that issue is the perceptual phenomenon that is reliable when it occurs'.[69] It is not advisable to 'force . . . observations and try to accommodate them to one's theories and opinions . . . looking for confirmation to theory rather than to the facts of observation'.[70] Nor is it advisable to 'transcend sense perception and to disregard it on the ground that "one ought to follow the argument"'.[71] The best course is 'to adopt the method already mentioned, to begin with phenomena . . . and, when this is done, to proceed to state the causes of those phenomena and to deal with their development'.

The methodological requirements are combined with a theory of perception that makes them plausible and gives them force. 'The sensitive and cognitive faculties of the soul' says Aristotle[72] 'are potentially these objects viz. the sensible and the knowable. The faculties, then, must be identified either with the objects themselves, or with their forms. Now they are not identical with the objects; for the stone does not exist in the soul, but only the form of the stone.' 'The sentient subject, as we have

---

[65] What follows is not a historical account of the Aristotelian opus but a philosophical account of what can be done with it (and has been done with it, e.g. by St. Thomas Aquinas).

[66] *An Post.* 100a3ff., esp. 100b2.

[67] *An. Pr.* 46a17ff.

[68] *An. Post.* 81a38.

[69] *De coelo* 306a7, Owens' translation.

[70] *De coelo* 293a27 – sounds very much like Newton!

[71] *de gen. et corr.* 325a13 – against Parmenides.

[72] *de anima* 431b26ff.

said, is potentially such as the object of sense is actually. Thus during the process of being acted upon it is unlike, but at the end of the process it has become like that object, and shares its quality.'[73] '*That which sees does in a sense possess colour*;[74] for each sense organ is receptive of the perceived object, but without its matter. This is why, even when the objects of perception are gone, sensations and mental images are still present in the sense organ'.[75] In the act of perception the very forms of nature are present in the mind, not merely images of them. Going against perception therefore means going against nature itself. Following perception means giving a true account of nature.[76] Aristotle's general theory of change which aided science down to the 19th century,[77] and which was supported by evidence of the most convincing kind, makes such an account most plausible.

On the other hand, it is not asserted that every single act of perception agrees with nature. The Aristotelian theory describes what happens during perception *in the normal case*. But the normal case may be distorted and even entirely concealed by disturbances. 'Error . . . seems to be more natural for living beings, and the soul spends more time on it'.[78] The disturbances must be studied and removed before knowledge can be obtained.

We have seen that the process by which universals are 'established' in the soul depends on particulars and 'low level universals' already imprinted in it. An idiosyncratic history of perception will therefore lead to idiosyncratic perceptions later on. Also the senses, being acquainted with our everyday surroundings, are liable to give misleading reports of objects outside this domain. This is proved by the appearance of the sun and the moon; on earth large but distant objects in familiar surroundings such as mountains are seen as being large, but far away. The moon and

[73] 418a2ff.

[74] There is, however, a difference between the way in which a property arises in a sense organ and in a physical body. Heating a physical body involves destruction of coldness in it. Producing the sensation of heat means actualizing a potentiality without destruction (cf. *de anima* 417b2ff. as well as Brentano, *Die Psychologie des Aristoteles*, Mainz 1867, p. 81). The reason for the difference is that a sense is not simply a physical body but a relation between extreme states (424a6f.).

[75] 425b23ff.

[76] This is not 'induction'. There is no 'inference' from 'evidence' to something different from it for the 'evidence' is already the thing sought for.

[77] Aristotle's law of inertia, for example (things remain in their state unless disturbed from the outside) which is repeated by Descartes, *Princ. Phil.*, section 37, has aided biologists in their research down to the beginning of this century (discovery of insect eggs, bacteria, viruses etc.). Newton's law would have been completely useless in these domains.

[78] *de an.* 427b3ff.

the sun, however, 'appear to measure one foot across even to men who are in health, and know [their] real measurements'.[79] The discrepancy is due to the imagination which is 'some kind of movement . . . caused by actual sensation'.[80] The movement 'resides in us and resembles sensations'[81] but 'it may be false . . . especially when the sensible object' appears in unusual conditions, such as large distance[82] and removed from supervision by the 'controlling sense'.[83] A combination of unusual conditions and absence of control thus leads to illusions; for example, patterns on the wall are sometimes seen as animals.[84]

Reading such passages we see that Aristotle was aware of the difficulties of astronomical observation,[85] he knew that the senses, used under exceptional circumstances can give exceptional and erroneous reports, he knew how to explain such reports and so he would not at all have been surprised by the problems of the first telescopic observations. Compared with him the 'modern' observers and especially Galileo approached the matter with a great and quite naive confidence. Ignorant of the psychological problems of telescopic vision, unfamiliar with the physical laws governing light in a telescope they went ahead and changed our world view. This was seen very clearly by Ronchi and some of his followers.[86]

Apart from unusual conditions error may also be due to the reaction of the senses themselves,[87] it may come from processes such as imagination which are triggered by sensations,[88] from 'mistakes in the operation of nature' comparable to 'monstrosities' in biology,[89] it may occur because the senses have been overstrained – 'the excitation is too strong, . . . the ratio of the adjustment (between sense and surroundings) is destroyed'[90]

[79] *de somn.* 458b28; cf. *de an.* 428b4ff.

[80] *de an.* 428b12ff.

[81] 429a5.

[82] 428b30f.

[83] *de somn.* 460b17.

[84] 460b12. Cf. also *Met.* 1010b14 on the perception of objects that are 'foreign' or 'strange' to the sense perceiving them as well as *de part. animal* 644b25 where it is said that the objects of astronomy 'though excellent and beyond compare and divine, are less accessible to knowledge. The evidence that might throw light on them and the problems concerning them is but scantily furnished by observation' and so errors are likely to arise.

[85] This awareness may be the reason why he never revised the homocentric system and never even mentioned its observational difficulties. For these difficulties and their later use cf. *AM*, Appendix 1.

[86] The problem of telescopic observation is discussed in chapters 10ff. of *AM*.

[87] *de somn.* 460b24.

[88] *de anima* 428a10.

[89] *Phys.* 199a38.

[90] *de an.* 425b25.

or when emotion, illness, large distance or other unusual conditions interfere with the proper working of the senses.[91] There are subliminal stimuli[92] which produce large scale actions of the organism affected[93] and there are imperceptible events[94] which still have their effects. Objects not appropriate to the sense by which they are perceived are more likely to lead to error than the proper objects of the sense concerned (colour, in the case of sight[95]), but even here we can have 'mistakes in the operations of nature' as we have seen. Misguided by such events we may become inclined to believe a false theory as 'based on experience' and may be forced to reject it *'because one can see no reasonable cause why it should be so'*[96]: Aristotle is prepared to 'squar[e] a recalcitrant fact with an empirical hypothesis'.[97] All this refutes Randall's assertion that Aristotle 'had no sense of the possibility of correction by more accurate means of observation'.[98] It also shows that the empiricism of Aristotle was more sophistica-

---

[91] *de somn.* 460b11 – the examples used here and their explanations show that Aristotle could have given a perfectly acceptable account of the strange phenomena reported in the first telescopic observations.

[92] *de divin. per somn.* 463a8.

[93] 463a29.

[94] *Meteor.* 355b20.

[95] *Met.* 1010b14; *de anima* 428b18.

[96] *de divin. per somn.* 462b14.

[97] G. E. L. Owen in *Aristotle*, ed. Moravcik, New York 1967, p. 171

[98] *Aristotle*, New York 1960, p. 57.

In his *Objective Knowledge*, Oxford 1973, p. 8, Popper writes with characteristic modesty: 'Neither Hume nor any other writer on the subject before me has to my knowledge moved from here (impossibility of justifying reasoning from experienced to unexperienced instances) to the *further questions*: can we take the "experienced instances" for granted? And are they really prior to the theories?'

It is surprising and a sign both of the historical illiteracy of most contemporary philosophers and of their low standards of hero worship that statements such as these are taken as historical evidence and as indications of philosophical profundity. But Newton corrected phenomena 'from above' [Cf. my 'Classical Empiricism' in J. W. Davis and R. E. Butts (eds.), *The Methodological Heritage of Newton*, Blackwell, Oxford 1970], Mill requested that there be a discussion of experience in order to determine both its content and its force ['On Liberty', in Marshall Cohen (ed.), *The Philosophy of John Stuart Mill*, New York 1961, p. 208] in Goethe's *Maximen and Reflexionen* we find the statement 'Das Höchste wäre zu begreifen, dass alles Faktische schon Theorie ist' (*Aus den   anderjahren*), expressing the very same idea which Popper here claims for himself. Boltzmann has often quoted Goethe's dictum that experience is only half experience [*Populäre Schriften*, Leipzig 1905, p. 222] and then there is, of course, Mach's observation that already the 'name "sensations" entails a one-sided theory' [*Analyse der Empfindungen*, Jena 1900, p. 8: 'Da aber in diesem Namen (der Empfindung) schon eine *einseitige* Theorie liegt . . .' emphasis in the original]. All this was unknown to the followers of the Vienna Circle who wanted to start philosophy afresh, and who did start it afresh, and with only minimal knowledge of earlier ideas. The Vienna Circle shares with the enlightenment an exaggerated faith in the *powers* of reason and an almost total ignorance concerning its *past achievements* – small wonder that Popper who anxiously

ted than either his critics or even some of his followers seemed to realize.

The difference between Aristotelian empiricism and the empiricism implicit in modern science (as opposed to the empiricism that turns up in the more philosophical pronouncements of scientists) thus does not lie in the fact that the former overlooks observational error while the latter is aware of it. *The difference lies in the role which error is permitted to play.* In Aristotle error beclouds and distorts particular perceptions *while leaving the general features of perceptual knowledge untouched.* However great the error, these general features can always be restored and it is from them that we receive information about the world we inhabit.[99] Aristotelian philosophy corresponds to commonsense. Commonsense, too, admits error, it has found ways of dealing with it, some forms of science included, but it will never concede that it is false throughout. Error is a *local phenomenon*, it does not distort our *entire outlook*. Modern science, on the other hand (and the Platonic and Democritian philosophies it absorbed) postulated just such *global distortions*. When it arose in the 16th and 17th centuries, it 'call[ed] in question a whole system, not just a particular detail; moreover, it was an attack not only on the physicists but on almost all sciences and all received opinions . . .'[100]

In *AM* and in earlier sections of the present book I discussed some aspects of this global change. I pointed out that the *arguments* that contributed to it were effective only because certain *changes of attitude* had also taken place. Such changes were partly the results of further arguments, partly non-intellectual reactions to new historical circumstances. 'Aristotle' lost followers and anti-Aristotelian diatribes succeeded because of the incompetence of many Aristotelians, because of new religious tendencies that revived the earlier clash between Aristotle and Christianity,[101] because of a rising revulsion against authority,[102] because

---

hovered at its periphery regarded every modification of the philosophy of the Vienna Circle as a genuine discovery. In this he was a true representative of the Vienna neo-enlightenment. But he did write a two-volume work on Plato, Aristotle, Hegel and other unfortunates and so one might expect him to have familiarized himself with their philosophies. Did he notice that the empiricist Aristotle put precisely the question, the 'further question' for which he now claims prime authorship? Apparently not. Which does not prevent him from criticizing Aristotle for his 'lack of insight' (*Open Society*, Vol. 2, p. 2).

[99] To this corresponds the one-by-one treatment of difficulties commented upon in n. 56.

[100] *Physics* 253a31ff. – speaking about Parmenides. To this corresponds the tendency to perceive anomalies as exhibiting a pattern and the attempt to gain entirely new views from them. Cf. n. 53.

[101] Cf. n. 59 and text.

[102] The elements varied from place to place so that a proper account will have to analyse the one 'Copernican Revolution' into many different but related processes of thought. Cf. also Francis R. Johnson, *Astronomical Thought in Renaissance England*, Baltimore 1937.

of the belief that there might be an America of Knowledge just as there was a new geographical continent, 'America'. They succeeded because of the support they received from philosophical and religious-mystical views and some occasionally rather unscientific ideas about man and the world. There was the belief in the infinite perfectibility of man and a corresponding distrust of commonsense. Both soul and body were assumed to be rich and changeable, capable of being influenced by training, instruments, learning of (old and) new things. It would be of great interest to know to what extent attitudes such as these which have been studied for quite some time [103] enlarged well-known technical difficulties (changes in the 8th sphere; problem of the comets; discovery of the vacuum, of the rough surface of the moon, of the moons of Jupiter) and turned them from *puzzles* to be solved within the old framework into *anticipations* of a new world. Some of the difficulties were old hat. Plutarch and then Oresme had argued for the rough surface of the moon, Oresme in exactly the same manner as Gilileo, but the argument became effective only in the 17th century – an excellent demonstration that arguments without attitudes achieve nothing. There were also *new methodological standards*. The Aristotelian philosophy could incorporate new ideas in two ways. It could either absorb them into its basic assumptions or use them as instruments of prediction ('saving phenomena'). There was never any change in the basic philosophy. Now the demand for 'novel predictions' became important. One is not satisfied with descriptions that *perfectly fit* the state of affairs described and with theories that can *accommodate* such descriptions, one also wants to 'become wiser', one wants to move beyond the horizon of the things known and knowable. We find this new demand in some later criticisms of Aristotle. These criticisms do not produce *arguments* against commonsense and, for philosophies that go beyond it, they simply *assume* that such philosophies are better and *revile* Aristotle for not being up to their standards. We have already seen how Copernicus and some of his followers preferred mathematical harmony to 'physics' by which they meant Aristotle.[104] Combining mathematics with the mechanical hypothesis Leibnitz criticized the Aristotelians for failing to give explanations. 'I should like you to think of one thing' he writes in his letter to Conring of March 19, 1687[105] 'that unless physical things can be explained by mechanical

---

[103] For details cf. e.g. P. O. Kristeller, *Renaissance Thought*, Vols. 1 and 2, Harper Torchbooks.

[104] Cf. above, n. 53 and text.

[105] Quoted from L. E. Loemker ed., *Gottfried Wilhelm Leibnitz, Philosophical Papers and Letters*, Dordrecht 1969, p. 189. My emphasis.

laws, God cannot, even if he chooses, reveal and explain nature to us. Or what would he say, I ask you, about vision and light? That light is the action of a potentially transparent body?[106] *Nothing is truer even though it is almost too true.* But would this make us any wiser? Could we use this to explain why the angle of reflection of light is equal to the angle of incidence, or why a ray should be bent more towards the perpendicular in a denser transparent body, though it would seem that the opposite should happen? . . . how can we hope to explain the causes of such things except by mechanical laws that is, by concrete mathematics or geometry applied to motion?'

The need for transcending commonsense is clearly *alluded to* in this diatribe but the only thing resembling an *argument* is a rhetorical question involving the words 'wise' and 'explain'. Now in (Aristotelian) commonsense 'explanation' has a good sense though it does not include anything like mechanical models. It is therefore *insinuated*, though not *stated* that science needs depth. But *this* demand is answered and rejected in Aristotle's arguments against Parmenides and the atomists. And Aristotle's theory of mathematics has solved problems which derive precisely from the assumption that the mathematical continuum has a 'depth structure'. Leibnitz' rhetorical question therefore does not advance matters a little bit.[107]

Now if one compares the difference between Aristotle and the ideology of modern science as expressed in Leibnitz with the arguments of Section 3 one notices an underlying difference of cosmologies. Aristotle's cosmos is finite, both qualitatively and quantitatively (though there is the *possibility* e.g. of infinite subdivision), it is viewed by an observer who can

---

[106] This is Aristotle's definition in *de anima* 418b9f.

[107] Today the situation is essentially the same except that the philosophy defended has become the *status quo* and has lots of problems (also Leibnitz was a few orders of magnitude more intelligent than his modern imitators). To get a whiff of the quality of 'modern' arguments let us take a look at what some 'critical rationalists' have written about the matter. There is the remark that the Aristotelian way of incorporating new facts involves 'degenerating problem shifts'. To start with, this is not an argument, it is a simple statement. The statement consists of two parts, viz. a *description* of what is done, using special terminology ('degenerating' etc.) and an *evaluation* of the events and procedures so described. It is interesting to see that the description *insinuates* the evaluation (who is ready to praise 'degeneration'?) and thereby creates norm–fact confusion, a matter which Popperians deplore whenever it occurs in the rhetoric of their opponents but cultivate when they need it for rhetorical displays of their own. But let us now overlook this quite legitimate attempt to forge *words* into weapons for knocking out an opponent and let us ask what reasons there are for the *evaluation*: why is the Aristotelian procedure unacceptable? No reply is forthcoming. Imre Lakatos rejects Aristotle because his philosophy does not agree with the standards of the methodology of research programmes. And how are the standards obtained?

grasp its basic structure if left in his normal state, and whose capabilities are fixed and also finite. The observer may use mathematics and other conceptual and physical artifices – but these have no ontological implications. The cosmos of modern science is an infinite world, mathematically structured, comprehended by the mind though not always by the senses and viewed by an observer whose abilities change from one discovery to the next. There is no stable equilibrium between man and the world though there are periods of stasis when the observer can settle, for a few decades, in a temporary home. The Aristotelian philosophy fits the first case, modern science and its philosophy the second.[108] So, one question is: what kind of world do we live in?

There is a second question; it is rarely raised by intellectuals but is quite important, much more important than the first question and it came to the fore only recently, after science had taken over almost all parts of public life and much of private life. It is this. Assume man has ingredients that can be uncovered, one by one, by progressive research of the second type, using mathematics and models from physics, chemistry, microbiology. Should we proceed and uncover them? And, having uncovered them, should we now see man in their light? Or would such a procedure not replace persons by the ahuman constituents of humanity and make us see everything in terms of the latter? And if that is the case – would it not be better to let research and realistic description stop at the common-sense level and to regard the remaining elements as complex instruments of prediction? Especially in view of the fact that the micro-account omits or overlooks global relations which are essential for our view of others and which play a large (and quite successful) role in non-scientific systems of medicine (the backwardness of much 'scientific' cancer research is due to a neglect of precisely such relations). Similar questions arise about the relation between man and nature. Here again (the awareness of) global

---

From the science of the 'last two centuries': Aristotle is rejected because his philosophy is not the philosophy of modern science. But that is the point at issue (for details cf. Chapter 16 of *AM* as well as my article in C. Howson (ed.) *Method and Appraisal in the Physical Sciences*, Cambridge 1976). Popper himself has nothing to contribute to the matter. He develops a methodology which is supposed to reflect modern science and which he uses against all other forms of knowledge. But in order to find an argument against Aristotle he would have to find difficulties in Aristotle that are *independent* of the fact that Aristotle does not use the methods of modern science. No such difficulties are ever mentioned. Thus the 'argument' boils down to: Aristotle is not like us – to hell with Aristotle! Typical critical rationalism!

[108] What I call the 'second case' is of course something much more recent than the mechanical philosophy of Descartes-Leibnitz-Newton which soon was turned into another *system*.

relations that linked man to nature have been destroyed with disastrous effects. Man once possessed complex knowledge concerning his place in nature and was to that extent secure and free. The knowledge has been replaced by abstract theories he does not understand and must take on trust from experts. But should humans not be able to understand the basic constituents of their lives? Should not every group, every tradition be able to influence, revere, preserve such constituents in accordance with its wishes? Is not the present separation of experts and sheep one reason for the much deplored social and psychological imbalance? And is it therefore not important to revive a philosophy that draws a clear distinction between a *natural knowledge* that is accessible to all and guides them in their relations to nature and their fellow men and the intellectual tumours, also called 'knowledge' that have assembled around it and have almost made it disappear?

Considerations such as these find support in new developments concerning the first question. Some interpreters of the quantum theory have pointed out that there is a natural limit beyond which mathematics ceases to reflect the world and becomes an instrument for the ordering of facts and this natural limit is provided by commonsense as improved (though not basically changed) by classical mechanics. For Heisenberg[109] this means a partial return to Aristotelian ideas. We know that tribal medicine, folk medicine, traditional forms of medicine in China, which remain close to a commonsense view of man and nature, have often better means of diagnosis and therapy than scientific medicine. We also know that 'primitive' forms of life have solved problems of human existence that are inaccessible to 'rational' treatment.[110] New developments in systems theory emphasize global relations using all the instruments of modern science but with the socio-natural properties and functions of man firmly in mind. All these developments must be considered when trying to come to a decision about Aristotle. What is needed for such a decision are again not only arguments, but also a new attitude, a new view of man and nature, a new religion that gives the arguments force just as new cosmology was needed to give force to the arguments of the Copernicans.

Which brings me to the last point of this section. In the preceding section I have asked three questions:

    A. Are there rules and standards which are 'rational' in the sense that

---

[109] *Physics and Philosophy*, New York 1964.
[110] Cf. E. Jantsch, *Design for Evolution*, New York 1975.

they agree with some plausible general principles and demand attention under all circumstances, which are obeyed by all good scientists when doing good research and whose adoption explains events such as the 'Copernican Revolution'?

B. Was it reasonable, at a given time, to accept Copernicus and what were the reasons? Did the reasons vary from one group to the next? From one period to the next?

C. Was there a time when it became unreasonable to reject Copernicus? Or is there always a point of view that permits us to regard the idea of a motionless earth as a reasonable idea?

I think it is clear both from Section 2 and from Sections 5 and 6 that the answer to A must be no. This, of course, was also the conclusion of *AM*.

The answer to B is yes with the proviso that different arguments convinced different people endowed with different attitudes. Maestlin liked mathematics and so did Kepler. Both were impressed by the harmony of the Copernican world system. Gilbert, having examined the motions of magnets was ready to adopt the motion of the earth as well. Guericke was impressed by the physical properties of the new system, Bruno by the fact that it could easily become part of an infinity of systems. Not one reason, not one method, but a variety of reasons made active by a variety of attitudes created the 'Copernican Revolution'. The reasons and attitudes converged but the convergence was accidental and it is vain trying to explain the whole process by the effects of simplistic methodological rules.

To answer question C we have to remember how Copernicus started. In the beginning his view was as unreasonable as the idea of the unmoved earth must have been in 1700. But it led to developments we now want to accept. Hence, it was reasonable to introduce it and try to keep it alive. Hence, it is always reasonable to introduce and try to keep alive unreasonable views.

7. Incommensurability

In Section 2 we have found how some traditions conceal structural elements beneath apparently accidental features while others exhibit them for everyone to see but hide the machinery that turns the structure into a language and an account of reality. We have also found that philosophers have often regarded traditions of the first kind as gross material to be shaped by traditions of the second kind. Proceeding in this way they confounded easy access with the presence and difficulty of discovery with the absence of structural elements and made the further mistake of assuming that structural elements that are formulated explicitly are the only operative ingredients of a language. The mistake is the main reason why philosophers of science have been content with discussing formulae and simple rules and why they have believed that such a discussion will eventually reveal everything that needs to be known about scientific theories. It was *Wittgenstein's* great merit to have spotted and criticized this procedure and the mistake that underlies it and to have emphasized that science contains not only formulae and rules for their application but entire *traditions*. *Kuhn* has expanded the criticism and made it more concrete. A *paradigm*, for him, is a tradition containing easily identifiable features side by side with tendencies and procedures that are not known but guide research in a subterranean way and are discovered only by contrast with other traditions. Introducing the notion of a paradigm Kuhn stated above all a *problem*. He explained to us that science depends on circumstances that are not described in the usual accounts, do not occur in science textbooks, and have to be identified in a roundabout way. Most of his followers, especially in the social sciences, did not see this problem but regarded Kuhn's account as the presentation of a new clear *fact* viz. the fact the word 'paradigm' identifies. Using a term awaiting explication by research as if the explication had already been completed they started a new and most deplorable trend of loquacious illiteracy (it is different with Lakatos who tries to identify some of the relevant features). In the present section I shall make a few brief comments on the notion of incommensurability which is a natural consequence of identifying theories with traditions and I shall also explain some differences between Kuhn's notion of incommensurability and mine.[111]

Kuhn has observed that different paradigms (A) use *concepts* that cannot be brought into the usual logical relations of inclusion, exclusion, overlap; (B) make us see things differently (research workers in different paradigms have not only different concepts, but also different *perceptions*[112]); and, (C) contain different *methods* (intellectual as well as physical instruments of research) for setting up research and evaluating its results. It was a great advance to replace the bloodless notion of a *theory* that had so far dominated discussions in the philosophy of science

---

[111] Further remarks about incommensurability are found in Part Three, Chapter 3, n. 38.

[112] This part was argued with vigour and many examples by the late N-R. Hanson, *Patterns of Discovery*, Cambridge 1958.

by the much more complex and subtle notion of a *paradigm* which one might call a theory-in-action and which included some of the dynamic aspects of science. According to Kuhn the collaboration of elements A, B and C makes paradigms fairly immune to difficulties and incomparable to each other.[113]

As opposed to Kuhn my own research started from certain problems in area A and concerned theories only.[114] Both in my thesis (1951) and in my first English paper on the matter[115] I asked how observation statements should be interpreted. I rejected two accounts viz. the 'pragmatic theory' according to which the meaning of an observation statement is determined by its use and the 'phenomenological theory' according to which it is determined by the phenomenon that makes us assert it as true. I interpreted observation languages by the theories that explain what we observe. Such interpretations change as soon as the theories change.[116] I realized that interpretations of this kind might make it impossible to establish deductive relations between rival theories and I tried to find means of comparison that were independent of such relations.[117] In the years following my 1958 paper (which preceded Kuhn's *Structure* and appeared in the same year as Hanson's *Patterns*) I tried to specify the conditions under which two theories 'in the same domain' would be deductively disjoint.[118] I also tried to find methods of comparison that

[113] Prof. Stegmüller, using certain methods of Snee'd has tried to reconstruct Kuhn's idea of a paradigm, of paradigm changes, of incommensurability but he did not succeed. Cf. my review in *BJPS*, Dec. 1977.

[114] Originally, under the influence of Wittgenstein I considered things very similar to paradigms ('language games'; 'forms of life' were the terms I used then) and I regarded them as comprising elements from A, B as well as C: different language games with different rules would give rise to different concepts, different ways of statement evaluation, different perceptions and would therefore be incomparable. I explained such ideas in Anscombe's house in Oxford in Fall 1952 with Hart and von Wright present. 'Making a discovery' I said 'often is not like finding America, but like waking up from a dream'. Later on I found it necessary to restrict research to be able to make more specific assertions. Kuhn's book and especially Lakatos' reactions to it then encouraged me to resume the more general approach. The results are found in Chapters 16 and 17 of *AM*. But much to the disgust of my colleagues in the philosophy of science I never used the narrow notion of 'theory'. Cf. my explanation in n. 5 of 'Reply to Criticism', *Boston Studies in the Philosophy of Science*, Vol. ii, New York 1965.

[115] 'An attempt at a Realistic Interpretation of Experience', *Proc. Arist. Soc.*, 1958, pp. 143ff. Published in German with a historical appendix in *Der wissenschaftstheoretische Realismus und die Autorität der Wissenschaften*, Vieweg Wiesbaden 1978.

[116] p. 163.

[117] Thus in my 1958 paper I tried to give an interpretation of crucial experiments that was independent of shared meanings. I improved this account in *Criticism and the Growth of Knowledge*, p. 226.

[118] The conditions deal only with theories and their logical relations and thus belong to

survive despite the absence of deductive relations. Thus while Ik was the incomparability of paradigms that results from the collaboration of A, B and C my version, If is deductive disjointness and nothing else and I never inferred incomparability from it. Quite the contrary, I tried to find means of comparing such theories. Comparison by *content*, or *verisimilitude* was of course out. But there certainly remained other methods.[119]

Now the interesting thing about these 'other methods' is that most of them, though reasonable in the sense that they agree with the wishes of a sizeable number of researchers are arbitrary, or 'subjective', in the sense that it is very difficult to find wish-independent arguments for their acceptability.[120] Also, these 'other methods' most of the time give conflicting results: a theory may be preferable because it makes numerous predictions, but the predictions may be based on rather daring approximations. On the other hand a theory may seem attractive because of its coherence but this 'inner harmony' may make it impossible to apply it to

---

area A of the paradigm differences noted by Kuhn. I believed for some time that conceptual differences would always be accompanied by perceptual differences but I abandoned this idea in 'Reply to Criticism', *op. cit.*, text to footnotes 50ff. Reason: the idea does not agree with results of psychological research. In *AM*, p. 238ff. I already warned against 'an inference from style (or language) to cosmology and mode of perception' and specified conditions in which such an inference can be made. To circumvent the difficulty that arises when we want to say that incommensurable theories 'speak about the same things' I restricted the discussion to non-instantial theories (*Minnesota Studies*, Vol. iii, 1962, p. 28) and I emphasized that mere *difference* of concepts does not suffice to make theories incommensurable in my sense. The situation must be rigged in such a way that the conditions of concept formation in one theory forbid the formation of the basic concepts of the other (cf. the explanation in *AM*, p. 269 and the reason, given there, why such explanations have to remain vague; cf. also the comparison of theory changes that lead to incommensurability with changes that do not in 'On the "Meaning" of Scientific Terms', *Journ. Philos.* 1965, Section 2). Of course, theories may be interpreted in different ways, they may be incommensurable in some interpretations, not incommensurable in others. Still, there are pairs of theories which in their customary interpretation turn out to be incommensurable in the sense at issue here. Examples are classical physics and quantum theory; general relativity and classical mechanics; Homeric Aggregate physics and the substance physics of the Presocratics.

[119] There are *formal criteria*: a linear theory is preferable to a non linear one because solutions can be obtained more easily. This was one of the main arguments against the nonlinear electrodynamics of Mie, Born and Infeld. The argument was also used against the general theory of relativity until the development of high speed computers simplified numerical calculations. Or: a 'coherent' theory is preferable to a non-coherent one (this was one of Einstein's reasons for prefering general relativity to other accounts). A theory using many and daring approximations to reach 'its facts' may be less likeable than a theory that uses only few, and safe approximations. Number of facts predicted may be another criterion. *Nonformal criteria* usually demand conformity with basic theory (relativistic invariance; agreement with basic quantum laws) or with metaphysical principles (Such as Einstein's 'principle of reality').

[120] Take simplicity, or coherence: why should a coherent theory be preferable to a non-coherent one? It is more difficult to handle, derivations of predictions are usually more

results in widely differing domains. Transition to criteria not involving content thus turns theory choice from a 'rational' and 'objective' routine into a complex decision involving conflicting preferences and propaganda will play a major role in it, as it does in all cases involving arbitrary elements.[121] Adding areas (B) and (C) strengthens the subjective, or 'personal' component of theory change.

To avoid such consequences the champions of objectivity and content increase have devised interpretations that turn incommensurable theories into commensurable ones. They overlook that the interpretations they so blithely push aside were introduced to solve a variety of physical problems and that incommensurability was just a side effect of these solutions. Thus the standard interpretation of the quantum theory was designed to explain penetration of potential barriers, interference, conservation laws, Compton effect, the photoelectric effect in a coherent way. And one important interpretation of the theory of relativity was introduced to make it independent of classical ideas. It is not too difficult to dream up interpretations which make incommensurable theories commensurable but not a single philosopher has so far been able to let his interpretation solve all the problems solved by the interpretation it is supposed to replace. In most cases these problems are not even known. Also philosophers have so far hardly dealt with areas B and C. Most of the time they simply assume that theory change leaves methods unchanged. Matters of perception are not even considered. Here Kuhn is far ahead of all positivists.

Incommensurability also shows that a certain form of realism is both too narrow and in conflict with scientific practice. Positivists believed that science deals essentially with observations. It orders and classifies observations, it never goes beyond them. Scientific change is a change of classificatory schemes blown up by a mistaken reification of the schemes. The critics of positivism pointed out that the world contains much more than observations. There are organisms, fields, continents, elementary particles, murders, devils and so on. Science, according to the critics, gradually discovers these things, determines their properties and their mutual relations. It makes the discoveries without changing the objects, properties, relations discovered. This is the essence of the realist position.

_____

elaborate and if the devil is master of this earth and a foe of scientists (why he should be I cannot imagine – but let us assume he is) then he will try to confound them so that simplicity and coherence will no longer be reliable guides.

[121] The issue between coherence on the one side, closeness to experimental results on the other, played a large role in the debates about the interpretation of the quantum theory.

Now realism may be interpreted as a *particular theory* about the relation between man and the world, and it may be interpreted as a *presupposition of science* (and knowledge in general). It seems that most philosophical realists adopt the second alternative – they are dogmatists. But even the first alternative can now be criticized and shown to be incorrect. All we need to do is to point out how often the world changed because of a change in basic theory. If the theories are commensurable, then no problem arises – we simply have an addition to knowledge. It is different with incommensurable theories. For we certainly cannot assume that two incommensurable theories deal with one and the same objective state of affairs (to make the assumption we would have to assume that both at least *refer* to the same objective situation. But how can we assert that 'they both' refer to the same situation when 'they both' never make sense together? Besides, statements about what does and what does not refer can be checked only if the things referred to are described properly, but then our problem arises again with renewed force.) Hence, unless we want to assume that they deal with nothing at all we must admit that they deal with different worlds and that the change (from one world to another) has been brought about by a switch from one theory to another. Of course, we cannot say that the switch was *caused* by the change (though matters are not quite as simple as that: waking up brings new principles of order into play and thereby causes us to perceive a waking world instead of a dream world). But since Bohr's analysis of the case of Einstein, Podolsky and Rosen we know that there are changes which are not results of a causal interaction between object and observer but of a change of the very conditions that permit us to speak of objects, situations, events. We appeal to changes of the latter kind when saying that a change of universal principles brings about a change of the entire world. Speaking in this manner we no longer assume an objective world that remains unaffected by our epistemic activities, except when moving within the confines of a particular point of view. We concede that our epistemic activities may have a decisive influence even upon the most solid piece of cosmological furniture – they may make gods disappear and replace them by heaps of atoms in empty space.[122]

---

[122] For a more detailed account cf. Gonzalo Muné Var, *Radical Knowledge*, Dissertation, Berkeley 1975.

Part Two
Science in a Free Society

1. Two Questions

There are two questions that arise in the course of any discussion of science. They are:

(A) *What is science?* – how does it proceed, what are its results, how do its standards, procedures, results differ from the standards, procedures, results of other fields?

(B) *What's so great about science?* – what makes science preferable to other forms of existence, using different standards and getting different results as a consequence? What makes modern science preferable to the science of the Aristotelians, or to the cosmology of the Hopi?

Note that in trying to answer question (B) we are not permitted to judge the alternatives to science by scientific standards. When trying to answer question (B) we *examine* such standards, so we cannot make them the basis of our judgements.

Question A has not one answer, but many. Every school in the philosophy of science gives a different account of what science is and how it works. In addition there are the accounts given by scientists, politicians and by so-called spokesmen of the general public. We are not far from the truth when saying that the nature of science is still shrouded in darkness. Still, the matter is discussed and there is a chance that some modest knowledge about science will some day arise.

There exists hardly anyone who asks question B. The excellence of science is *assumed*, it is not *argued for*. Here scientists and philosophers of science act like the defenders of the One and Only Roman Church acted before them: Church doctrine is true, everything else is Pagan nonsense. Indeed, certain methods of discussion and insinuation that were once treasures of theological rhetoric have now found a new home in science.

This phenomenon, though remarkable and somewhat depressing, would hardly bother a sensible person if it were restricted to a small number of the faithful: in a free society there is room for many strange beliefs, doctrines, institutions. But the assumption of the inherent superiority of science has moved beyond science and has become an article of faith for almost everyone. Moreover, science is no longer a particular institution; it is now part of the basic fabric of democracy just as the Church was once part of the basic fabric of society. Of course, Church and State are now carefully separated. State and Science, however, work closely together.

Immense sums are spent on the improvement of scientific ideas. Bastard subjects such as the philosophy of science which shares with science the name but hardly anything else profit from the boom of the sciences. Human relations are subjected to scientific treatment as is shown by education programmes, proposals for prison reform, army training and so on. The power of the medical profession over every stage of our lives already exceeds the power once wielded by the Church. Almost all scientific subjects are compulsory subjects in our schools. While the parents of a six-year-old can decide to have him instructed in the rudiments of Protestantism, or in the rudiments of the Jewish faith, or to omit religious instruction altogether, they do not have similar freedom in the case of the sciences. Physics, astronomy, history *must* be learned; they cannot be replaced by magic, astrology, or by a study of legends.

Nor is one content with a merely *historical* presentation of physical (astronomical, biological, sociological etc.) facts and principles. One does not say: *some people believe* that the earth moves around the sun while others regard the earth as a hollow sphere that contains the sun, the planets, the fixed stars. One says: the earth *moves* round the sun – everything else is nonsense.

Finally, the manner in which we accept or reject scientific ideas is radically different from democratic decision procedures. We accept scientific laws and facts, we teach them in our schools, we make them the basis of important political decisions, but without having examined them, and without having subjected them to a vote. *Scientists* do not subject them to a vote, or at least this is what they tell us, and laymen certainly do not subject them to a vote. Concrete proposals are occasionally discussed, and a vote is suggested (nuclear reactor initiatives). But the procedure is not extended to general theories and scientific facts. Modern society is 'Copernican' not because Copernicus was put up for vote, discussed in a democratic way, and voted in with a simple majority; it is 'Copernican' because the *scientists* are Copernicans and because one accepts their cosmology as uncritically as one once accepted the cosmology of bishops and of cardinals.

Even bold and revolutionary thinkers bow to the judgement of science. Kropotkin wants to break up all existing institutions, but he does not touch science. Ibsen goes very far in his critique of bourgeois society, but he retains science as a measure of truth. Lévi Strauss has made us realize that Western thought is not the lonely peak of human achievement it was once thought to be, but he and his followers exclude science from their relativization of ideologies.[1] Marx and Engels were convinced that science would aid the workers in their quest for mental and social liberation.

Such an attitude made perfect sense in the 17th, 18th, even 19th centuries when science was one of many competing ideologies, when the state had not yet declared in its favour and when its determined pursuit was more than balanced by alternative views and alternative institutions. In those years science was a liberating force, not because it had found the truth, or the right method (though this was assumed to be *the* reason by the defenders of science), but because it restricted the influence of other ideologies and thus gave the individual room for thought. Nor was it necessary in those years to press a consideration of question B. The opponents of science who still were very much alive tried to show that science was on the wrong track, they belittled its importance and the scientists had to reply to the challenge. The methods and achievements of science were subjected to a critical debate. In this situation it made perfect sense to commit oneself to the cause of science. The very circumstances in which the commitment took place turned it into a liberating force.

It does not follow that the commitment has a liberating effect today. There is nothing in science or in any other ideology that makes them inherently liberating. Ideologies can deteriorate and become dogmatic religions (example: Marxism). They start deteriorating when they become successful, they turn into dogmas the moment the opposition is crushed: their triumph is their downfall. The development of science in the 19th and 20th centuries and especially after the Second World War is a good example. The very same enterprise that once gave man the ideas and the strength to free himself from the fears and the prejudices of a tyrannical religion now turns him into a slave of its interests. And let us not be deceived by the libertarian rhetoric and by the great show of

---

[1] Lévi Strauss (*The Savage Mind*, Chicago 1966, pp. 16ff.) denies that myth, being 'the product of man's "mythmaking faculty" turn[s] its back on reality'. He sees in it an approach to nature that complements science and is characterized by a 'universe of instruments [that is] closed' while the scientist will try new procedures to get new results. There can never be a conflict between the results of science and myth and so the question of their relative merit can

tolerance that some propagandists of science are putting on for our benefit. Let us ask whether they would be prepared to give, say the views of the Hopi the same role in basic education which science has today, let us ask a member of the AMA whether he would permit faithhealers into state hospitals and we shall soon see how narrow the limits of this tolerance really are. And, mind you, these limits are not the results of research; they are imposed quite arbitrarily as we shall see later on.

2. The Prevalence of Science a Threat to Democracy

This symbiosis of the state and of an unexamined science leads to an interesting problem for intellectuals and especially for liberals.

Liberal intellectuals are among the chief defenders of democracy and freedom. Loudly and persistently they proclaim and defend freedom of thought, speech, religion and, occasionally, some quite inane forms of political action.

Liberal intellectuals are also 'rationalists'. And they regard rationalism (which for them coincides with science) not just as one view among many, but as a basis for society. The freedom they defend is therefore granted under conditions that are no longer subjected to it. It is granted only to those who have already accepted part of the rationalist (i.e. scientific) ideology.[2]

For a long time this dogmatic element of liberalism was hardly noticed, let alone commented upon. There are various reasons for the oversight. When Blacks, Indians and other suppressed races first emerged into the broad daylight of civic life their leaders and their supporters among Whites demanded equality. But equality, 'racial' equality included, then did not mean *equality of traditions*; it meant *equality of access to one particular tradition* – the tradition of the White Man. The Whites who supported the demand opened the Promised Land – but it was a Promised Land built after their own specifications and furnished with their own favourite playthings.

The situation soon changed. An increasing number of individuals and

never arise. Things look different to some Marxist critics. Thus M. Godelier ('Myth et Histoire', *Annales* 1971) lets myth transform the 'numerous objective data about nature into an "imaginative" explanation of reality' where 'objective data' are the data of science. Science, once more, has the upper hand.

[2] See n. 14, p. 29.

groups became critical of the gifts offered.[3] They either revived their own traditions or adopted traditions different both from rationalism and from the traditions of their forefathers. At this stage intellectuals started developing 'interpretations'. After all, they had studied non-Western tribes and cultures for quite some time. Many descendants of non-Western societies owe whatever knowledge they have of their ancestors to the work of white missionaries, adventurers, anthropologists, some of them with a liberal turn of mind.[4] When later anthropologists collected and systematized this knowledge they transformed it in an interesting way. The emphasized the psychological meaning, the social functions, the existential temper of a culture, they disregarded its ontological implications. According to them oracles, rain dances, the treatment of mind and body *express* the needs of the members of a society, they *function* as a social glue, they *reveal* basic structures of thought, they may even lead to an increased *awareness* of the relations between man and man and man and nature but without an accompanying *knowledge* of distant events, rain, mind, body. Such interpretations were hardly ever the result of critical thought – most of the time they were simply a consequence of popular antimetaphysical tendencies combined with a firm belief in the excellence first, of Christianity and then of science. This is how intellectuals, Marxists included aided by the forces of a society that is democratic in words only almost succeeded in having it both ways: they could pose as understanding friends of non-Western cultures without endangering the supremacy of their own religion: science.

The situation changed again. There are now individuals, some very gifted and imaginative scientists among them who are interested in a genuine revival not just of the externals of non-scientific forms of life but of the world views and practices (navigation, medicine, theory of life and matter) that were once connected with them. There are societies such as mainland China where traditional procedures have been combined with

---

[3] White middle class Christians (and liberals, rationalists, even Marxists) felt great satisfaction when they finally offered Indians some of the marvellous opportunities of the great society they think they inhabit and they were displeased and offended when the reaction was disappointment, not abject gratitude. But why should an Indian who never even dreamt of imposing his culture on a white man now be grateful for having white culture imposed on him? Why should he be grateful to the white man who, having stolen his material possessions, his land, his living space now proceeds to steal his mind as well?

[4] Christian missionaries occasionally had a better grasp of the inherent rationality of 'barbaric' forms of life than their scientific successors and they were also greater humanitarians. As an example the reader should consult the work of Las Casas as described in Lewis Hanke *All Mankind is One*, Northern Illinois Press 1974.

scientific views leading to a better understanding of nature and a better treatment of individual and social dysfunction. And with this the hidden dogmatism of our modern friends of freedom becomes revealed: democratic principles as they are practised today are incompatible with the undisturbed existence, development, growth of special cultures. A rational-liberal (-Marxist) society cannot contain a Black culture in the full sense of the word. It cannot contain a Jewish culture in the full sense of the word. It cannot contain a mediaeval culture in the full sense of the word. It can contain these cultures only as secondary grafts on a basic structure that is an unholy alliance of science, rationalism (and capitalism).[5]

But – so the impatient believer in rationalism and science is liable to exclaim – is this procedure not justified? Is there not a tremendous difference between science on the one side, religion, magic, myth on the other? Is this difference not so large and so obvious that it is unnecessary to point it out and silly to deny it? Does the difference not consist in the fact that magic, religion, mythical world views *try* to get in touch with reality while science *has succeeded* in this business and so supersedes its ancestors? Is it therefore not only justified but also required to remove an ontologically potent religion, a myth that claims to describe the world, a system of magic that poses as an alternative to science from the centre of society and to replace them by science? These are some of the questions which the 'educated' liberal (and the 'educated' Marxist) will use to object to any form of freedom that interferes with the central position of science and (liberal or Marxist) rationalism.

Three assumptions are contained in these rhetorical questions.

Assumption A: scientific rationalism is preferable to alternative traditions.

Assumption B: it cannot be improved by a comparison and/or combination with alternative traditions.

Assumption C: it must be accepted, made a basis of society and education because of its advantages.

---

[5] Professor Agassi, see Part Three, Chapter One, has read this passage as suggesting that Jews *should* return to the traditions of their forefathers, that American Indians *should* resume their old ways, rain dances included, and he has commented on the 'reactionary' character of such suggestions. Reactionary? This assumes that the step into science and technology was not a mistake – which is the question at issue. It also assumes, for example, that rain dances don't work – but who has examined that matter? Besides, I do not make the suggestion Agassi ascribes to me. I don't say that American Indians (for example) *should resume* their old ways, I say that those who *want to resume them* should be able to do so first, because in a democracy everyone should be able to live as he sees fit and second, because no ideology and no way of life is so perfect that it cannot learn from a comparison with alternatives.

In what follows I shall try to show that neither assumption A nor assumption B agrees with the facts where 'facts' are defined in accordance with the type of rationalism implicit in A and B: *rationalists and scientists cannot rationally (scientifically) argue for the unique position of their favourite ideology.*

However, assume they can – does it follow that their ideology must now be imposed on everyone (question C)? Is it not rather the case that traditions that give substance to the lives of people must be given equal rights and equal access to key positions in society *no matter what other traditions think about them*? Must we not demand that ideas and procedures that give substance to the lives of people be made full members of a free society *no matter what other traditions think about them*?

There are many people who regard such questions as an invitation to *relativism*. Reformulating them in their own favourite terms they ask us whether we would want to give falsehood the same rights as truth, or whether we would want dreams to be treated as seriously as accounts of reality. From the very beginning of Western Civilization insinuations such as these were used to defend one view, one procedure, one way of thinking and acting to the exclusion of everything else.[6] So, let us take the bull by its horns and let us take a closer look at this frightful monster: relativism.

3. The Spectre of Relativism

With the discussion of relativism we enter territory full of treacherous paths, traps, footangles, territory where appeals to emotion count as arguments and where arguments are of a touching simplemindedness. Relativism is often attacked not because one has found a fault, but because one is afraid of it. Intellectuals are afraid of it because relativism threatens their role in society just as the enlightenment once threatened

---

[6] In Plutarch's *Life of Solon* we find the following story: 'When the company of Thespis began to exhibit tragedy, and its novelty was attracting the populace but had not yet gone as far as public competition, Solon being fond of listening and learning and being rather given in his old age to leisure and amusements, and indeed to drinking parties and music, went to see Thespis act in his own play, as was the practice of ancient times. Solon approached him after the performance and asked him if he was not ashamed to tell so many lies to so many people. When Thespis said there was nothing dreadful in representing such works and actions in fun, Solon struck the ground violently with his walking stick: "If we applaud these things in fun" he said "we shall soon find ourselves honouring them in earnest". Thus began the 'long standing quarrel between poetry and philosophy' (Plato *Republic* 607b6f.), i.e. between those seeing everything in terms of truth and falsehood, and other traditions.

the existence of priests and theologians. And the general public which is educated, exploited and tyrannized by intellectuals has learned long ago to identify relativism with cultural (social) decay. This is how relativism was attacked in Germany's Third Reich, this is how it is attacked again today by Fascists, Marxists, Critical Rationalists. Even the most tolerant people dare not say that they reject an idea or a way of life because they don't like it – which would put the blame on them entirely – they have to add that there are *objective* reasons for their action – which puts at least part of the blame on the thing rejected and on those enamoured by it. What is it about relativism that seems to put the fear of god into everyone?

It is the realization that one's own most cherished point of view may turn out to be just one of many ways of arranging life, important for those brought up in the corresponding tradition, utterly uninteresting and perhaps even a hindrance to others. Only few people are content with being able to think and live in a way pleasing to themselves and would not dream of making their tradition an obligation for everyone. For the great majority – and that includes Christians, rationalists, liberals and a good many Marxists – there exists only one truth and it must prevail. Tolerance does not mean acceptance of falsehood side by side with truth, it means human treatment of those unfortunately caught in falsehood.[7] Relativism would put an end to this comfortable exercise in superiority – therefore the aversion.

Fear of moral and political chaos increases the aversion by adding practical disadvantages to the intellectual drawbacks. Relativists, it is said, have no reason to respect the laws of the society to which they belong, they have no reason to keep promises, honour business contracts, respect the lives of others, they are like beasts following the whim of the moment and like beasts they constitute a danger to civilized life.

It is interesting to see how closely this account mirrors the complaints of Christians who witnessed the gradual removal of *religion* from the centre of society. The fears, insinuations and predictions were then exactly the same – but they did not come true. Replacing religion by rationalism and science did not create paradise – far from it – but it did not create chaos either.

It did not create chaos, it is pointed out, because rationalism is itself an orderly philosophy. One order was replaced by another order. But relativism wants to remove *all* ideological ingredients (except those that are convenient, for the time being). Is it possible to have such a society?

[7] Cf. Henry Kamen, *The Rise of Toleration*, New York 1967.

Can it work? How will it work? These are the questions we have to answer.

Starting with the intellectual (or semantic) difficulties viz. the insinuation that relativism means giving the same rights to truth and falsehood (reason and insanity, virtue and viciousness and so on) we need only remind the reader of theses i. and ii. of Section 2, Part One and the associated explanations. We saw then that classifying traditions as true or false (. . . etc. . . . .) means projecting the point of view of other traditions upon them. Traditions are neither good nor bad – they just are. They obtain desirable or undesirable properties only for an agent who participates in another tradition and projects the values of this tradition upon the world. The projections *appear 'objective'* i.e. tradition-independent and the statements expressing its judgements *sound 'objective'* because the subject and the tradition he represents nowhere occur in them. They *are* '*subjective*' because this non-occurrence is due to an oversight. The oversight is revealed when the agent adopts another tradition: his value-judgements change. Trying to account for the change the agent has to revise the content of all his value statements just as physicists had to revise the content of even the simplest statement about length when it was discovered that length depends on the reference system. Those who don't carry out the revision cannot pride themselves on forming a special school of especially astute philosophers who have withstood the onslaught of moral relativism just as those who still cling to absolute lengths cannot pride themselves on forming a special school of especially astute physicists who have withstood the onslaught of relativity. They are just pig-headed, or badly informed, or both. So much about seeing relativism in terms of equal rights for falsehood, irrationality, viciousness and so on.

That the appeal to truth and rationality is rhetorical and without objective content becomes clear from the inarticulateness of its defence. In Section 1 we have seen that the question 'What is so great about science?' is hardly ever asked and has no satisfactory answer. The same is true of other basic concepts.[8] Philosophers inquire into the nature of truth, or the nature of knowledge, but they hardly ever ask why truth should be pursued (the question arises only at the boundary line of traditions – for example, it arose at the boundary line of science and Christianity). The very same notions of Truth, Rationality, Reality that are supposed to eliminate relativism are surrounded by a vast area of

---

[8] Can I use 'truth' when criticizing its uncritical use? Of course I can, just as one can use German to explain the disadvantages of German and the advantages of Latin to a German audience.

ignorance (which corresponds to the arguer's ignorance of the tradition that provides the material for his rhetorical displays).

There is therefore hardly any difference between the members of a 'primitive' tribe who defend their laws because they are the laws of their gods, or of their ancestors and who spread these laws in the name of the tribe and a rationalist who appeals to 'objective' standards, except that the former know what they are doing while the latter does not.[9]

This concludes the intellectual, or 'semantic' part of the debate about relativism.

Turning now to the political problems we can start by pointing out that many of them are entirely imaginary. The assumption that they plague only relativists and resist solution except within the framework of a particular tradition (Christianity, Rationalism) is simply slander – aided by insufficient analysis. For we must distinguish between political relativism and philosophical relativism and we must separate the psychological attitude of relativists from both. *Political relativism* affirms that all traditions have equal *rights*: the mere fact that some people have arranged their lives in accordance with a certain tradition suffices to provide this tradition with all the basic rights of the society in which it occurs.

---

[9] The rules of a rational science, liberal intellectuals say, do not involve special interests. They are 'objective' in the sense that they emphasize truth, reason etc. all of which are independent of the beliefs and wishes of special interest groups. Distinguishing between the *validity* of a demand, a rule, a suggestion and the fact that the demand, rule, suggestion is *accepted* critical rationalists seem to turn knowledge and morals from tribal ideologies into the representation of tribe-independent circumstances. But tribal ideologies do not cease to be tribal ideologies on account of not being openly characterized as such. The demands which rationalists defend and the notions they use *speak* 'objectively' and not in the name of Sir Karl Popper or Professor Gerard Radnitzky because *they have been made to speak that way* and not because the interests of Sir Karl or of Professor Radnitzky are no longer taken into account; and they have been made to speak that way to secure them a wider audience, to keep up the pretence of libertarianism and because rationalists have little sense for what one might call the 'existential' qualities of life. Their 'objectivity' is in no way different from the 'objectivity' of a colonial official who, having read a book or two now ceases to address the natives in the name of the King and addresses them in the name of Reason instead or from the 'objectivity' of a drill sergeant who instead of shouting 'now, you dogs, listen to me – this is what I want you to do and God have mercy on you if you don't do exactly what I tell you!' purrs 'Well, I think what we ought to do is . . .'. Obedience to the commands and the ideology of the speaker is demanded in either case. The situation becomes even clearer when we examine how rationalists argue. They posit a 'truth' and 'objective' methods for finding it. If the necessary concepts and methods are known to all the parties in the debate, then nothing further needs to be said. The debate can start right away. If one party does not know the methods, or uses methods of its own then it must be *educated* which means *it is not taken seriously* unless its procedure coincides with the procedure of the rationalist. Arguments are tribe-centred and the rationalist is the master.

A 'more philosophical' argument might support such a procedure by pointing out that traditions are neither good nor bad but simply are (Part One, Section 2, Thesis 1), that they assume positive or negative qualities only when viewed through the spectacles of other traditions (Thesis ii) and that the judgement of those who live in accordance with the tradition is to be given preference. *Philosophical relativism* is the doctrine that all traditions, theories, ideas are equally true or equally false or, in an even more radical formulation, that any distribution of truth values over traditions is acceptable. This form of relativism is nowhere defended in the present book. It is not asserted, for example, that Aristotle is as good as Einstein, it is asserted and argued that 'Aristotle is true' is a judgement that presupposes a certain tradition, it is a relational judgement that *may* change when the underlying tradition is changed. There *may* exist a tradition for which Aristotle is as true as Einstein, but there are other traditions for which Einstein is too uninteresting for examination. Value judgements are not 'objective' and cannot be used to push aside the 'subjective' opinions that emerge from different traditions. I also argue that the appearance of objectivity that is attached to some value judgements comes from the fact that a particular tradition is *used* but not *recognized*: absence of the impression of subjectivity is not proof of 'objectivity' but of an oversight.

Turning now to the *attitudes* of relativists we must distinguish between (a) members of a relativistic society and (b) philosophical relativists. Among the former we shall find all attitudes from sheer dogmatism combined with a strong urge to proselytize to an out-and-out liberalism/cynicism. Political relativism makes assertions about *rights* (and about protective structures defending these rights) – not about beliefs, attitudes etc. Philosophical relativists, on the other side, may again have all sorts of attitudes, punctilious obedience to the law included.

Now one seems to assume that acceptance of political relativism will drastically increase the number of those who only want to please themselves and that everybody will be subjected to their whims. I regard this assumption as most implausible. Only few of the traditions of a relativistic society will be lawless – most of them will regiment their members even more strongly than is done in the so-called 'civilized societies' of today. The assumption also insinuates that it is lack of indoctrination and not lack of choice that is responsible for the drastic increase of the crime rate we observe today so that it is not fear of retaliation but the proper education that makes people behave decently – a wildly implausible theory. Christianity preached love for mankind and burned, killed, maimed hundreds of thousands of people. The French Revolution preached Reason and Virtue and ended up in an ocean of blood. The USA were built on the right to liberty and the pursuit of happiness for all – and yet there was slavery, suppression, intimidation. One could of course insist that the failure was due to inefficient methods of education – but 'more efficient' methods would be neither wise nor humane. Eradicate the ability to kill – and people may lose their passion. Eradicate the ability to lie – and imagination which always goes against the truth of the moment might disappear as well (cf. n. 6). An 'education' in virtue might easily make people incapable of being wicked by making them incapable of being people – a large price to pay for results that can be achieved in other ways. And that there are such other ways is openly admitted by the anti-relativists. Far from trusting the force of the ideology whose importance they emphasize with such passion they protect society by laws, courts, prisons, and an efficient police force. But a police force can be used by relativists as well, for – and with this we come to the second part of the assumption at the beginning of this paragraph – such a society will not be and cannot be without protective devices. It is to be admitted that speaking of police, prisons, protection does not sound good in the ears of those concerned with freedom. However a universal training in virtue and rationality that obliterates traditions and is liable to create meek zombies is an even greater threat to it. What kind of protection is better – the inefficient protection that comes from interfering with the soul or the much more efficient *external* protection that *leaves souls intact* and only restricts our movements?

A relativistic society will therefore contain a *basic protective structure*. This leads to the next argument for rationalism (or some similar central protective ideology): must not the structure be 'just'? Must it not be shielded from undue influence? Must there not be an 'objective' way of settling disputes about it which means – is there not again a need for rationalism over and above particular traditions?

To answer this question we need only realize that protective frameworks are not introduced out of the blue but in a concrete historical situation and that it is this situation and not an abstract discussion of 'justice' or 'rationality' that determines the process. People living in a society that does not give their tradition the rights they think it deserves will work towards a change. To effect the change, they will use the most efficient means available. They will use existing laws, if that is going to help their cause, they will 'argue rationally' when rational argument is required, they will engage in an open debate (cf. the explanations to

Part One, Section 2, Thesis viii.) where the representatives of the status quo have no fixed opinion and no fixed procedure, they will organize an uprising if there seems no other way. To demand that they restrict their efforts to what is rationally admissible may at that stage be as sensible as the demand to reason with a wall. Besides, why should they worry about 'objectivity' when their aim is to make themselves heard in one-sided, i.e. 'subjective', surroundings?

The situation is different when tribes, cultures, people who are not part of any one state move into the same area and now have to live together. An example are Babylonians, Egyptians, Greeks, Mitanni, Hittites and the many other peoples who had interests in Asia Minor. They learned from each other and created the 'First Internationalism' (Brestead) of 1600 to 1200 B.C. Tolerance of different traditions and different creeds was considerable and by far exceeded the tolerance which Christians later showed towards alternative forms of life. The Yassaq of Genghis Khan which proclaims the same rights for all religons shows that history does not always progress and that the 'modern mind' may be far behind some 'savages' as regards reasonableness, practicality and tolerance.

The third case is that of a relativistic society with a protective structure already installed. This is the case which rationalists seem to have in mind. We want to improve the protective structure. The improvement, rationalists say, must not be done arbitrarily, there must not be undue influence, objective standards must determine every single step. But why should the standards that guide an exchange between traditions be imposed from the outside? In Part One we have seen that the relation between Reason and Practice is a dialectical relation: traditions are guided by standards which are in turn judged by the way in which they influence them. The same is true of the standards that guide the exchange between the various traditions of a free society. These standards are again determined, improved, refined, eliminated by the traditions themselves or, to use terms explained in the same place – *the exchange between traditions is an open exchange, not a rational exchange.* Insinuating that the internal business of a society must follow 'objective' rules, pointing out that they are the foremost inventors, guardians, polishers of rules, intellectuals have so far succeeded in interposing themselves between the traditions concerned and their problems. They have succeeded in preventing a more direct democracy where problems are solved and solutions judged by those who suffer from the problems and have to live with the solutions and they have fattened themselves on the funds thus diverted in their direction. It is time to realize that they are just one special and rather greedy group held together by a special and rather aggressive tradition equal in rights to Christians, Taoists, Cannibals, Black Muslims but often lacking their understanding of humanitarian issues. It is time to realize that science, too, is a special tradition and that its predominance must be reversed by an open debate in which all members of the society participate.

But – and with this we proceed to question A of Section 2 – will such a debate not soon discover the overwhelming excellence of science and thus restore the status quo? And if it doesn't – does this not show the ignorance and incompetence of laymen? And if that is so, is it then not better to leave things as they are instead of disturbing them by useless and time-consuming changes?

4. Democratic Judgement overrules 'Truth' and Expert Opinion

There are two aspects to this question. One concerns the *rights* of citizens and traditions in a free society, the other the (perhaps disadvantageous) consequences of an *exercise* of these rights.

In a democracy an individual citizen has the right to read, write, to make propaganda for whatever strikes his fancy. If he falls ill, he has the right to be treated in accordance with his wishes, by faithhealers, if he believes in the art of faithhealing, by scientific doctors, if he has greater confidence in science. And he has not only the right to accept, live in accordance with, and spread ideas *as an individual*, he can *form associations* which support his point of view provided he can finance them, or find people willing to give him financial support. This right is given to the citizen for two reasons; first, because everyone must be able to pursue what he *thinks* is truth, or the correct procedure; and, secondly, because the only way of arriving at a useful judgement of what is supposed to be the truth, or the correct procedure is to become acquainted with the widest possible range of alternatives. The reasons were explained by Mill in his immortal essay *On Liberty*. It is not possible to improve upon his arguments.

Assuming this right, a citizen has a say in the running of any institution to which he makes a financial contribution, either privately, or as a tax-payer: state colleges, state universities, tax supported research institutions such as the National Science Foundation are subjected to the judgement of taxpayers, and so is every local elementary school. If the taxpayers of California want their state universities to teach Voodoo, folk medicine, astrology, rain dance ceremonies, then this is what the universities will have to teach. Expert opinion will of course be taken into consideration, but experts will not have the last word. The last word is the decision of democratically constituted committees, and in these committees laymen have the upper hand.

But do laymen possess the knowledge that is needed for decisions of this kind? Will they not commit grievous mistakes? And is it not therefore necessary to leave fundamental decisions to the experts?

Certainly not in a democracy.

A democracy is an assembly of mature people and not a collection of sheep guided by a small clique of know-it-alls. Maturity is not found lying about in the streets, it must be learned. It is not learned in schools, at least not in the schools of today where the student is confronted with desiccated and falsified *copies* of *past decisions*, it is learned by *active participation* in decisions that are still to be made. Maturity is more important than special knowledge and it must be pursued even if the pursuit should interfere with the delicate and refined charades of the scientists. After all, we have to decide how special forms of knowledge are to be applied, how far they may be trusted, what their relation is to the *totality* of human existence and, therefore, to other forms of knowledge. Scientists, of course, assume that there is nothing better than science. The citizens of a democracy cannot rest content with such a pious faith. Participation of laymen in fundamental decisions is therefore required *even if it should lower the success rate of the decisions.*

The situation I have just described has many similarities with the situation in a case of war. In a war a totalitarian state has a free hand. No humanitarian considerations restrict its tactics; the only restrictions are those of material, ingenuity, manpower. A democracy, on the other hand, is supposed to treat the enemy in a humane fashion *even if this should lower the chances of victory*. It is true that only few democracies ever live up to such standards but those that do make an important contribution to the advancement of our civilization. In the domain of thought the situation is exactly the same. We must realize that there are more important things in this world than winning a war, advancing science, finding the truth. Besides, it is not at all certain that taking fundamental decisions out of the hands of experts and leaving them to laymen is going to lower the success rate of the decisions.

5. Expert Opinion often Prejudiced, Untrustworthy, and in Need of Outside Control

To start with, experts often arrive at different results, both in fundamental matters, and in application. Who does not know of at least one case in his family where one doctor recommends a certain operation, another argues against it, while a third suggests an entirely different procedure? Who has not read of the debates about nuclear safety, the state of the economy, the effects of pesticides, aerosol sprays, the efficiency of methods of education, the influence of race on intelligence? Two, three, five and even more views arise in such debates, and scientific supporters can be found for all of them. Occasionally one almost feels inclined to say: as many scientists, as many opinions. There are of course areas in which scientists agree – but this cannot raise our confidence. Unanimity is often the result of a *political* decision: dissenters are suppressed, or remain silent to preserve the reputation of science as a source of trustworthy and almost infallible knowledge. On other occasions unanimity is the result of shared prejudices: positions are taken without detailed examination of the matter under review and are infused with the same authority that proceeds from detailed research. The attitude towards astrology which I shall discuss presently is an example. Then again unanimity may indicate a decrease of critical consciousness: criticism remains faint as long as only one view is being considered. This is the reason why a unanimity that rests on 'internal' considerations alone often turns out to be mistaken.

Such mistakes *can be* discovered by laymen and dilettantes, and often *have been* discovered by them. Inventors built 'impossible' machines and made 'impossible' discoveries. Science was advanced by outsiders, or by scientists with an unusual background. Einstein, Bohr, Born were dilettantes, and said so on numerous occasions. Schliemann who refuted the idea that myth and legend have no factual content started as a successful businessman, Alexander Marshack who refuted the idea that Stone Age man was incapable of complex thought as a journalist, Robert Ardrey was a playwright and came to anthropology because of his belief in the close relation between science and poetry, Columbus had no university education and had to learn Latin late in his life, Robert Mayer knew just the bare outlines of early 19th century physics, the Chinese communists of the Fifties who forced traditional medicine back into the universities and thereby started most interesting lines of research the world over had only little knowledge of the intricacies of scientific medicine. How is this possible? How is it possible that the ignorant, or ill-informed can occasionally do better than those who know a subject inside out?

One answer is connected with the very *nature of knowledge*. Every piece of knowledge contains valuable ingredients side by side with ideas that prevent the discovery of new things. Such ideas are not simply errors. They are necessary for research: progress in one direction cannot be achieved without blocking progress in another. But research in that 'other' direction may reveal that the 'progress' achieved so far is but a chimera. It may seriously undermine the authority of the field as a whole. Thus science needs both the *narrowmindedness* that puts obstacles in the path of an unchained curiosity and the *ignorance* that either disregards the obstacles, or is incapable of perceiving them.[10] Science needs both the expert and the dilettante.[11]

Another answer is that scientists quite often just don't know what they are talking about. They have strong opinions, they know some standard arguments for these opinions, they may even know some results outside the particular field in which they are doing research but most of the time they depend, and have to depend (because of specialization) on *gossip* and *rumours*. No special intelligence, no technical knowledge is needed to find this out. Anyone with some perseverance can make the discovery and he will then also find that many of the rumours that are presented with such assurance are nothing but simple mistakes.

Thus R. A. Millikan, Nobel Prize Winner in Physics writes in *Reviews*

[10] Ignorance of established school doctrines helped Galileo in his research. Ignorance made others adopt the results of his research, despite grave observational and conceptual difficulties. This is shown in Chapters 9–11 and Appendix 2 of *AM*.

[11] It is interesting to see that the demands of the new experimental philosophy that appeared in the 17th century eliminated not just hypotheses, or methods, *but the very effects* whose spuriousness was afterwards said to have been proved by scientific research: parapsychological effects and effects showing a harmony between microcosm and macrocosm depend on a state of mind (and, in the case of large scale phenomena, of society) that is eliminated by the demand for 'unprejudiced and neutral observers'; these effects increase with excitement, a global approach and a close correlation of spiritual and material agencies. They decrease and almost disappear when a cool and analytical approach is taken, or when religion and theology are separated from the study of inert matter. Thus scientific empiricism eliminated its spiritualistic rivals, it eliminated the followers of Agrippa of Nettesheim, John Dee, Robert Fludd not by giving a better account of a world *that existed independently of either view*, but by using a method that did not permit 'spiritual' effects to arise. It *removed* such effects and then described the impoverished world *insinuating that no change had taken place*. James I who did not feel too comfortable with spirits could only welcome such a development and we have reasons to assume that 'scientists' craving for Royal patronage arranged their science accordingly. Bacon's changing attitude towards magic should be seen in this light also: cf. F. Yates, *The Rosicrucian Enlightenment*, London 1974.

of Modern Physics, Vol. 29 (1949), p. 344: 'Einstein called out to us all – "let us merely accept this (the Michelson experiment) as an established experimental fact and from there proceed to work out its inevitable consequences" – and he went at the task himself with an energy and a capacity which very few people on earth possess. Thus was born the special theory of relativity'.

The quotation suggests that Einstein starts with the description of an experiment, that he urges us to lay aside prior ideas and to concentrate on the experiment alone, that he himself abandons such ideas, and that using this method he arrives at the special theory of relativity. One has only to read Einstein's paper of 1905 to realize that he proceeds in an entirely different way. There is no mention of the Michelson–Morley experiment or, for that matter, of any particular experiment. The starting point of the argument is not an experiment, but a 'conjecture' and Einstein's suggestion is, not to eliminate the 'conjecture', but to 'raise (it) into a principle' – the very opposite of what Millikan describes Einstein as doing. This can be verified by anyone who is able to read, without special knowledge of physics, for the passage occurs in the first and non-mathematical part of Einstein's paper.

Another and more technical example is the so-called *Neumann proof*. In the Thirties there existed two major interpretations of the quantum theory. According to the first interpretation quantum theory is a statistical theory, like statistical mechanics, and the uncertainties are uncertainties of knowledge, not uncertainties of nature. According to the second interpretation the uncertainties do not merely express our ignorance, they are inherent in nature: states that are more definite than indicated by the uncertainty relations simply do not exist. The second interpretation was defended by Bohr who offered a variety of qualitative arguments and by Heisenberg who illustrated it with simple examples. In addition there was a somewhat complicated proof by von Neumann allegedly showing that quantum mechanics was incompatible with the first view. Now at meetings up to the Fifties the discussion usually went like this. First the defenders of the second interpretation presented their arguments. Then the opponents raised objections. The objections were occasionally quite formidable and could not be easily answered. Then somebody said 'but von Neumann has shown . . .' and with this the opposition was silenced and the second interpretation saved. It was saved not because von Neumann's proof was so well known but because the mere name 'von Neumann' was an authority to overrule any objection. It was saved because of the force of an authoritative *rumour*.

At this point the similarity between 'modern' science and the Middle Ages becomes rather striking. Who does not remember how objections were defused by reference to Aristotle? Who has not heard of the many rumours (such as the rumour that the young of a lion are born dead and licked to life by their mother) that were handed on from generation to generation and formed decisive parts of mediaeval knowledge? Who has not read with indignation how observations were rejected by reference to theories which were just further rumours and who has not either himself pontificated or heard others pontificate on the excellence of modern science in this respect? The examples show that the difference between modern science and 'mediaeval' science is at most a matter of degree and that the same phenomena occur in both. The similarity increases when we consider how scientific institutions try to impose their will on the rest of society.[12]

6. The Strange Case of Astrology

To drive the point home I shall briefly discuss the 'Statement of 186 leading Scientists' against astrology which appeared in the September/ October issue 1975 of the *Humanist*. This statement consists of four parts. First, there is the statement proper which takes about one page. Next come 186 signatures by astronomers, physicists, mathematicians, philosophers and individuals with unspecified professions, eighteen Nobel Prize Winners among them. Then we have two articles explaining the case against astrology in some detail.

Now what surprises the reader whose image of science has been formed by the customary eulogies which emphasize rationality, objectivity, impartiality and so on is the religious tone of the document, the illiteracy of the 'arguments' and the authoritarian manner in which the arguments are being presented. The learned gentlemen have strong convictions, they use their authority to spread these convictions (why 186 signatures if one has arguments?), they know a few phrases which sound like arguments, but they certainly do not know what they are talking about.[13]

[12] Numerous examples in Robert Jungk, *Der Atomstaat*, Munich 1977.

[13] This is quite literally true. When a representative of the BBC wanted to interview some of the Nobel Prize Winners they declined with the remark that they had never studied astrology and had no idea of its details. Which did not prevent them from cursing it in public. In the case of Velikowski the situation was exactly the same. Many of the scientists

Take the first sentence of the 'Statement'. It reads: 'Scientists in a variety of fields have become concerned about the increased acceptance of astrology in many parts of the world.'

In 1484 the Roman Catholic Church published the *Malleus Maleficarum*, the outstanding textbook on witchcraft. The *Malleus* is a very interesting book. It has four parts: phenomena, aetiology, legal aspects, theological aspects of witchcraft. The description of phenomena is sufficiently detailed to enable us to identify the mental disturbances that accompanied some cases. The aetiology is pluralistic, there is not just the official explanation, there are other explanations as well, purely materialistic explanations included. Of course, in the end only one of the offered explanations is accepted, but the alternatives are discussed and so one can judge the arguments that lead to their elimination. This feature makes the *Malleus* superior to almost every physics, biology, chemistry textbook of today. Even the theology is pluralistic, heretical views are not passed over in silence, nor are they ridiculed; they are described, examined, and removed by argument. The authors know the subject, they know their opponents, they give a correct account of the positions of their opponents, they argue against these positions and they use the best knowledge available at the time in their arguments.

The book has an introduction, a bull by Pope Innocent VIII, issued in 1484. The bull reads: 'It has indeed come to our ears, not without afflicting us with bitter sorrow, that in . . .' – and now comes a long list of countries and counties – 'many persons of both sexes, unmindful of their own salvation have strayed from the Catholic Faith and have abandoned themselves to devils . . .' and so on. The words are almost the same as the words in the beginning of the 'Statement', and so are the sentiments expressed. Both the Pope and the '186 leading scientists' deplore the increasing popularity of what they think are disreputable views. But what a difference in literacy and scholarship!

Comparing the *Malleus* with accounts of contemporary knowledge the reader can easily verify that the Pope and his learned authors knew what they were talking about. This cannot be said of our scientists. They neither know the subject they attack, astrology, nor those parts of their own science that undermine their attack.

---

who tried to prevent the publication of Velikowski's first book or who wrote against it once it had been published never read a page of it but relied on gossip or on newspaper accounts. This is a matter of record. Cf. de Grazia, *The Velikowski Affair*, New York 1966, as well as the essays in *Velikovsky Reconsidered*, New York 1976. As usual the greatest assurance goes hand in hand with the greatest ignorance.

Thus Professor Bok, in the first article that is attached to the statement writes as follows: 'All I can do is state clearly and unequivocally that modern concepts of astronomy and space physics give no support – better said, negative support – to the tenets of astrology' i.e. to the assumption that celestial events such as the positions of the planets, of the moon, of the sun influence human affairs. Now, 'modern concepts of astronomy and space physics' include large planetary plasmas and a solar atmosphere that extends far beyond the earth into space. The plasmas interact with the sun and with each other. The interaction leads to a dependence of solar activity on the relative positions of the planets. Watching the planets one can predict certain features of solar activity with great precision. Solar activity influences the quality of short wave radio signals hence fluctuations in this quality can be predicted from the position of the planets as well.[14]

Solar activity has a profound influence on life. This was known for a long time. What was not known was how delicate this influence really is. Variations in the electric potential of trees depend not only on the *gross* activity of the sun but on *individual flares* and therefore again on the positions of the planets.[15] Piccardi, in a series of investigations that covered more than thirty years found variations in the rate of standardized chemical reactions that could not be explained by laboratory or meteorological conditions. He and other workers in the field are inclined to believe 'that the phenomena observed are primarily related to changes of the structure of water used in the experiments'.[16] The chemical bond

[14] J. H. Nelson, *RCA Review*, Vol. 12 (1951), pp. 26ff.; *Electrical Engineering*, Vol. 71 (1952), pp. 421ff. Many of the scientific studies that are relevant for our case are described and indexed in Lyall Watson, *Supernature*, London 1973. Most of these studies have been neglected (without criticism) by orthodox scientific opinion.

[15] This was found by H. S. Burr. Reference in Watson, *op. cit.*

[16] S. W. Tromp, 'Possible Effects of Extra-Terrestrial Stimuli on Colloidal Systems and Living Organisms', *Proc. 5th Intern. Biometeorolog. Congress, Nordwijk 1972*, Tromp and Bouma (eds.), p. 243. The article contains a survey of the work initiated by Piccardi who started long range studies on the causes of certain non-reproducible physico-chemical processes in water. Some of the causes were related to solar eruptions, others to lunar parameters. Reference to such extra terrestrial stimuli is rare among environmental scientists and the corresponding problems are 'often forgotten or neglected' (p. 239). However, 'despite a certain resistance experienced among orthodox scientists, a clear breakthrough can be observed in recent years amongst the younger research workers' (p. 245). There are special research centres such as the *Biometeorological Research Center* in Leiden and the *Stanford Research Center* in Menlo Park, California which study what once was called the influence of the heavens upon the earth and have found correlations between organic and unorganic processes and lunar, solar, planetary parameters. Tromp's article contains a survey and a large bibliography. The *Biometeorological Research Center* issues periodic lists of publications (monographs, reports, publications in scientific journals). Part of the work done at the *Stanford Research Institute* and related institutions is reported in (ed.) John Mitchell *Psychic Exploration, A Challenge for Science*, New York 1974.

in water is about one tenth of the strength of average chemical bonds so that water is 'sensitive to extremely delicate influences and is capable of adapting itself to the most varying circumstances to a degree attained by no other liquid.'[17] It is quite possible that solar flares have to be included among these 'varying circumstances'[18] which would again lead to a dependence on planetary positions. Considering the role which water and organic colloids[19] play in life we may conjecture that 'it is by means of water and the aqueous system that the external forces are able to react on living organisms'.[20]

Just how sensitive organisms are has been shown in a series of papers by F. R. Brown. Oysters open and close their shells in accordance with the tides. They continue their activity when brought inland, in a dark container. Eventually they adapt their rhythm to the new location which means that they sense the very weak tides in an inland laboratory tank.[21] Brown also studied the metabolism of tubers and found a lunar period though the potatoes were kept at constant temperature, pressure, humidity, illumination: man's ability to keep conditions constant is smaller than the ability of a potato to pick up lunar rhythms[22] and Professor Bok's assertion that 'the walls of the delivery room shield us effectively from many known radiations' turns out to be just another case of a firm conviction based on ignorance.

The 'Statement' makes much of the fact that 'astrology was part and parcel of (the) magical world view' and the second article that is attached to it offers a 'final disproof' by showing that 'astrology arose from magic'. Where did the learned gentlemen get *this* information? As far as one can see there is not a single anthropologist among them and I am rather doubtful whether anyone is familiar with the more recent results of this discipline. What they do know are some *older* views from what one might call the 'Ptolemaic' period of anthropology when post-17th century Western man was supposed to be the sole possessor of sound knowledge, when field studies, archaeology and a more detailed examination of myth had not yet led to the discovery of the surprising knowledge possessed by ancient man as well as by modern 'Primitives' and when it was assumed

---

[17] G. Piccardi, *The Chemical Basis of Medical Climatology*, Springfield, Illinois 1962.

[18] Cf. G. R. M. Verfaillie, *Intern. Journ. Biometeorol.*, Vol. 13 (1969), pp. 113ff.

[19] Tromp, *loc. cit.*

[20] Piccardi, *loc. cit.*

[21] *Am. Journ. Physiol.*, Vol. 178 (1954), pp. 510ff.

[22] *Biol. Bull.*, Vol. 112 (1957), p. 285. The effect could also be due to synchronicity – cf. C. G. Jung, 'Synchronicity: An Acausal Connecting Principle', in *The Collected Works of C. G. Jung*, Vol. 8, London 1960, pp. 419ff.

that history consisted in a simple progression from more primitive to less primitive views. We see: the judgement of the '186 leading scientists' rests on an antediluvian anthropology, on ignorance of more recent results in their own fields (astronomy, biology, and the connection between the two) as well as on a failure to perceive the implications of results they do know. It shows the extent to which scientists are prepared to assert their authority even in areas in which they have no knowledge whatsoever.

There are many minor mistakes. 'Astrology', it is said 'was dealt a serious death blow' when Copernicus replaced the Ptolemaic system. Note the wonderful language: does the learned writer believe in the existence of 'death blows' that are not 'serious'? And as regards the content we can only say that the very opposite was true. Kepler, one of the foremost Copernicans used the new discoveries to improve astrology, he found new evidence for it, and he defended it against opponents.[23] There is a criticism of the dictum that the stars incline, but do not compel. The criticism overlooks that modern hereditary theory (for example) works with inclinations throughout. Some specific assertions that are part of astrology are criticized by quoting evidence that contradicts them; but every moderately interesting theory is always in conflict with numerous experimental results. Here astrology is similar to highly respected scientific research programmes. There is a longish quotation from a statement by psychologists. It says: 'Psychologists find no evidence that astrology is of any value whatsoever as an indicator of past, present, or future trends of one's personal life . . .'. Considering that astronomers and biologists have not found evidence *that is already published, and by researchers in their own fields*, this can hardly count as an argument. 'By offering the public the horoscope as a substitute for honest and sustained thinking, astrologers have been guilty of playing upon the human tendency to take easy rather than difficult paths' – but what about psychoanalysis, what about the reliance upon psychological tests which long ago have become a substitute for 'honest and sustained thinking' in the evaluation of people of all ages?[24] And as regards the magical origin of astrology one need only remark that science once was very closely connected with magic and must be rejected if astrology must be rejected on these grounds.

[23] Cf. Norbert Herz, *Keplers Astrologie*, Vienna 1895, as well as the relevant passages from Kepler's collected works. Kepler objects to tropical astrology, retains sidereal astrology, but only for mass phenomena such as wars, plagues etc.

[24] The objection from free will is not new; it was raised by the Church fathers. So was the twin objection.

The remarks should not be interpreted as an attempt to defend astrology *as it is practiced now* by the great majority of astrologists. Modern astrology is in many respects similar to early mediaeval astronomy: it inherited interesting and profound ideas, but it distorted them, and replaced them by caricatures more adapted to the limited understanding of its practitioners.[25] The caricatures are not used for research; there is no attempt to proceed into new domains and to enlarge our knowledge of extra-terrestrial influences; they simply serve as a reservoir of naive rules and phrases suited to impress the ignorant. Yet this is not the objection that is raised by our scientists. They do not criticize the air of stagnation that has been permitted to obscure the basic assumptions of astrology, they criticize these basic assumptions themselves and in the process turn their own subjects into caricatures. It is interesting to see how closely both parties approach each other in ignorance, conceit and the wish for easy power over minds.[26]

7. Laymen can and must supervise Science

These examples, which are not at all atypical,[27] show that it would not only be foolish *but downright irresponsible* to accept the judgement of scientists and physicians without further examination. If the matter is important, either to a small group or to society as a whole, *then this judgement must be subjected to the most painstaking scrutiny.* Duly elected committees of laymen must examine whether the theory of evolution is really as well established as biologists want us to believe, whether being established in their sense settles the matter, and whether it should replace other views in schools. They must examine the safety of nuclear reactors in each individual case and must be given access to *all* the relevant information. They must examine whether scientific medicine deserves the unique position of theoretical authority, access to funds, privileges of mutilation it enjoys today or whether non-scientific methods of healing are not frequently superior and they must encourage relevant comparisons: traditions of tribal medicine must be revived and practiced by those who prefer them partly because it is their wish, partly because we thus obtain some information about the efficiency of science (cf. also the remarks in Section 9 below). The committees must also examine whether

[25] On astrology see *AM* p. 100n.
[26] Cf. *AM* p. 208n.
[27] Further examples are given in *AM*.

peoples' minds are properly judged by psychological tests, what is to be said about prison reforms – and so on and so forth. In all cases the last word will not be that of the experts, but that of the people immediately concerned.[28]

That the errors of specialists can be discovered by ordinary people provided they are prepared to 'do some hard work' is the basic assumption of any trial by jury. The law demands that experts be cross-examined and that their testimony be subjected to the judgement of a jury. In making this demand it assumes that experts are human after all, that they make mistakes, even right in the centre of their specialty, that they try to cover up any source of uncertainty that might reduce the credibility of their ideas, that their expertise is not as inaccessible as they often insinuate. And it also assumes that a layman can acquire the knowledge necessary for understanding their procedures and finding their mistakes.

---

[28] Scientists, educators, physicians must be supervised when engaged in *public* jobs; but they must also be watched most carefully when called upon to solve the *problems of an individual*, or a family. Everyone knows that plumbers, carpenters, electricians cannot always be trusted and that it is wise to keep an eye on them. One starts by comparing different firms, chooses the one making the best suggestions and supervises every step of their work. The same applies to the so-called 'higher' professions: an individual who engages a lawyer, consults a meteorologist, asks for a foundation report on his house cannot take things for granted or he will find himself with a large bill and problems even greater than those for whose solution he called in the expert. All this is pretty well known. But there are some professions which still seem to be exempt from doubt. Many people trust a physician or an educator as they would have trusted a priest in earlier times. But doctors give incorrect diagnoses, prescribe harmful drugs, cut, X-ray, mutilate at the slightest provocation partly because they are incompetent, partly because they don't care and have so far been able to get away with murder, partly because the basic ideology of the medical profession which was formed in the aftermath of the scientific revolution can deal only with certain restricted aspects of the human organism but still tries to cover everything by the same method. Indeed, so large has the scandal of malpractice become that the physicians themselves are now advising their patients not to be content with a single diagnosis but to shop around and to supervise their treatment. Of course, second opinions should not be restricted to the medical profession for the problem may not be the incompetence of a single doctor, or of a group of doctors, the problem may be the *incompetence of scientific medicine as a whole*. Thus every patient must be the supervisor of his treatment just as every group of people and every tradition must be allowed to judge the projects which the government wants to carry out in their midst and must be able to reject those projects it does not regard as adequate.

In the case of educators the situation is still worse. For while it is possible to determine whether a *physical* treatment has been successful we have no ready means to determine the success of a mental treatment, of a so-called education. Reading, writing, arithmetic and knowledge of basic facts can be judged. But what shall we think of a training that turns people into second-hand existentialists or philosophers of science? What shall we think of the idiocies propagated by our sociologists and the atrocities regarded as 'critical productions' by our artists? They can palm off their ideas on us with impunity unless pupils start checking out their teachers just as patients have started checking out their doctors: the advice in all cases is to *use experts*, but never to *trust them* and certainly never to *rely on them* entirely.

This assumption is confirmed in trial after trial. Conceited and intimidating scholars, covered with honorary degrees, university chairs, presidencies of scientific societies are tripped up by a lawyer who has the talent to look through the most impressive piece of jargon and to expose the uncertainty, indefiniteness, the monumental ignorance behind the most dazzling display of omniscience: *science is not beyond the reach of the natural shrewdness of the human race.* I suggest that this shrewdness be applied to all important social matters which are now in the hands of experts.

8. Arguments from Methodology do not Establish the Excellence of Science

The considerations presented so far may be criticized by admitting that science, being a product of human effort has its *faults* but by adding that it is still *better* than alternative ways of acquiring knowledge. Science is superior for two reasons: it uses the correct *method* for getting results; and there are many *results* to prove the excellence of the method. Let us take a closer look at these reasons.

The answer to the first reason is simple: there is no 'scientific method'; there is no single procedure, or set of rules that underlies every piece of research and guarantees that it is 'scientific' and, therefore, trustworthy. Every project, every theory, every procedure has to be judged on its own merits and by standards adapted to the processes with which it deals. The idea of a universal and stable *method* that is an unchanging measure of adequacy and even the idea of a universal and stable *rationality* is as unrealistic as the idea of a universal and stable measuring instrument that measures any magnitude, no matter what the circumstances. Scientists revise their standards, their procedures, their criteria of rationality as they move along and enter new domains of research just as they revise and perhaps entirely replace their theories and their instruments as they move along and enter new domains of research. The main argument for this answer is historical: there is not a single rule, however plausible and however firmly grounded in logic and general philosophy that is not violated at some time or other. Such violations are not accidental events, they are not avoidable results of ignorance and inattention. Given the conditions in which they occurred they were necessary for progress, or for any other feature one might find desirable. Indeed, one of the most striking features of recent discussion in the history and philosophy of science is the realization that events such as the invention of atomism in antiquity, the Copernican Revolution, the rise of modern atomism (Dalton; kinetic theory; dispersion theory; stereochemistry; quantum theory), the gradual emergence of the wave theory of light occurred only because some thinkers either *decided* not to be bound by certain 'obvious' rules, or because they *unwittingly broke* them. Conversely, we can show that most of the rules which are today defended by scientists and philosophers of science as constituting a uniform 'scientific method' are either useless – they do not produce the results they are supposed to produce – or debilitating. Of course, we may one day find a rule that helps us through all difficulties just as we may one day find a theory that can explain everything in our world. Such a development is not likely, one might almost be inclined to say that it is logically impossible, but I would still not want to exclude it. The point is that the development *has not yet started*: *today* we have to do science without being able to rely on any well defined and stable 'scientific method'.

The remarks made so far do not mean that research is arbitrary and unguided. There are standards, but they come from the research process itself, not from abstract views of rationality. It needs ingenuity, tact, knowledge of details to come to an informed judgement of existing standards and to invent new ones just as it needs ingenuity, tact, knowledge of details to come to an informed judgement of existing theories and to invent new ones. More of this in Section 3 of Part One and Section 3 of Chapter 4 of Part Three.

There are writers who agree with the account given so far and still insist on a special treatment for science. Polányi, Kuhn and others object to the idea that science must conform to external standards and insist as I do that standards are developed and examined by the very same process of research they are supposed to judge. This process, they say, is a most delicate machinery. It has its own Reason and determines its own Rationality. And therefore, so they add, it must be left undisturbed. Scientists will succeed only if they are entirely research oriented, if they are allowed to pursue only those problems they regard as important and to use only procedures that seem efficient to them.

This ingenious defence of financial support without corresponding obligations cannot be maintained. To start with, research is not always successful and often produces monsters. Small mistakes, involving restricted areas, may perhaps be corrected from the inside, comprehensive mistakes involving the 'basic ideology' of the field can be and often were revealed only by outsiders or by scientists with an unusual personal history. Making use of new ideas these outsiders corrected the mistakes and so changed research in a fundamental way. Now what counts and what does not count as a mistake depends on the tradition that does the judging: for an analytical tradition (say, in medicine) the important thing is to find basic elements and to show how everything is built up from them. Lack of immediate success is a sign of the complexity of the problem and the need for more and more efficient research *of the same kind*. For a holistic tradition the important thing is to find large scale connections. Lack of immediate success of the analytic tradition is now a sign of its (partial) inadequacy and new research strategies may be suggested (this, incidentally, is roughly the situation in certain parts of cancer research). In the beginning the suggestions will be regarded as unwanted interference just as the mixing of astronomical and physical arguments was regarded as unwanted interference by the Aristotelian physicists of the 16th and 17th centuries. Which leads to a further criticism of the Kuhn-Polányi view: it assumes that the distinctions and separations implicit in a certain historical stage are unobjectionable and have to be maintained. But different research programmes were often united, or one subsumed under the other with a resulting change in competences. There is no reason why the research programme *science* should not be subsumed under the research programme *free society* and competences changed and redefined accordingly. The change is needed – the possibilities of freedom will not be exhausted without it – there is nothing inherent in science (except the wish of scientists to do their own thing at other people's expense) that forbids it; many scientific developments, though on a smaller scale, have been of exactly the same kind and, besides, an independent science has long ago been replaced by the *business* science which lives off society and strengthens its totalitarian tendencies. This disposes of the Polányi-Kuhn objection.

9. Nor is Science Preferable because of its Results

According to the second reason science deserves a special position because of its *results*.

This is an argument only if it can be shown (a) that no other view has ever produced anything comparable and (b) that the results of science are autonomous, they do not owe anything to non-scientific agencies. Neither assumption survives close scrutiny.

It is true that science has made marvellous contributions to our under-standing of the world and that this understanding has led to even more marvellous practical achievements. It is also true that most rivals of science have by now either disappeared, or have been changed so that a conflict with science (and therefore the possibility of results that differ from the results of science) no longer arises: religions have been 'de-mythologized' with the explicit purpose of making them acceptable to a scientific age, myths have been 'interpreted' in a manner that removed their ontological implications. Some features of this development are not at all surprising. Even in a fair competition one ideology often assembles successes and overtakes its rivals. This does not mean that the beaten rivals are without merit and that they have ceased to be capable of making a contribution to our knowledge, it only means that they have tem-porarily run out of steam. They may return and cause the defeat of their defeaters. The philosophy of atomism is an excellent example. It was introduced (in the West) in antiquity with the purpose of 'saving' macrophenomena such as the phenomenon of motion. It was overtaken by the dynamically more sophisticated philosophy of the Aristotelians, returned with the scientific revolution, was pushed back with the development of continuity theories, returned again late in the 19th cen-tury and was again restricted by complementarity. Or take the idea of the motion of the earth. It arose in antiquity, was defeated by the powerful arguments of the Aristotelians, regarded as an 'incredibly ridiculous' view by Ptolemy, and yet staged a triumphant comeback in the 17th century. What is true of theories is true of methods: knowledge was founded on speculation and logic, then Aristotle introduced a more empirical procedure which was replaced by the more mathematical methods of Descartes and Galileo which in turn was combined with a fairly radical empiricism by the members of the Copenhagen school. The lesson to be drawn from this historical sketch is that a temporary setback for an ideology (which is a bunch of theories combined with a method and a more general philosophical point of view) must not be taken as a reason for eliminating it.

Yet this is precisely what happened to older forms of science and to non-scientific points of view after the scientific revolution: they were eliminated, first from science itself, then from society until we arrive at the present situation where their survival is endangered not only by the general prejudice in favour of science, but by institutional means as well: science has now become part of the basic fabric of democracy, as we have seen. In these circumstances, is it surprising that science reigns supreme and is the only ideology known to have worthwhile results? It reigns supreme because some *past successes* have led to institutional measures (education; role of experts; role of power groups such as the AMA) that prevent a comeback of the rivals. Briefly, but not incorrectly: *today science prevails not because of its comparative merits, but because the show has been rigged in its favour*.

There is another element involved in this rigging mechanism, and we must not overlook it. I said above that ideologies may fall behind even in a fair competition. In the 16th and 17th centuries there was a fair competition (more or less) between ancient Western science and philosophy and the new scientific philosophy; there was never any fair competition between this entire complex of ideas and the myths, religions, procedures of non-Western societies. These myths, these religions, these procedures have disappeared or deteriorated not because science was better, but because *the apostles of science were the more determined conquerors*, because they *materially suppressed* the bearers of alternative cultures. There was no research. There was no 'objective' comparison of methods and achievements. There was colonization and suppression of the views of the tribes and nations colonized. These views were replaced, first, by the religion of brotherly love, and then by the religion of science. A few scientists studied tribal ideologies, but being prejudiced and insufficiently prepared they were unable to find any evidence of superiority or even of equality (not that they would have recognized such evidence had they found it). Again the superiority of science is the result not of research, or argument, it is the result of political, institutional, and even military pressures.

To see what happens when such pressures are removed or used against science we need only take a look at the history of traditional medicine in China.

China was one of the few countries that escaped Western intellectual domination down to the 19th century. Early in the 20th century a new generation, tired of the old traditions and the restrictions implicit in them and impressed by the material and intellectual superiority of the West imported science. Science soon pushed aside all traditional elements. Herbal medicine, acupuncture, moxibustion, the yin/yang duality, the theory of the chi were ridiculed and removed from schools and hospitals, Western medicine was regarded as the only sensible procedure. This was the attitude up to about 1954. Then the party, realizing the need for a political supervision of scientists ordered traditional medicine back into hospitals and universities. The order restored the free competition between science and traditional medicine. One now discovered that traditional medicine has methods of diagnosis and therapy that are superior to those of Western scientific medicine. Similar discoveries were made by those who compared tribal medicines with science. The lesson to be learned is that *non-scientific ideologies, practices, theories, traditions can become powerful rivals and can reveal major shortcomings of science if only they are given a fair chance to compete.* It is the task of the institutions of a free society to give them such a fair chance.[29] The excellence of science, however, can be asserted only *after* numerous comparisons with alternative points of view.

More recent research in anthropology, archaeology (and here especially in the flourishing subject of archaeoastronomy,[30] history of science, parapsychology[31] has shown that our ancestors and our 'primitive' contemporaries had highly developed cosmologies, medical theories, biological doctrines which are often more adequate and have better results than their Western competitors[32] and describe phenomena not accessible

---

[29] In the 15th, 16th and 17th centuries artisans emphasized the conflict between their concrete knowledge and the abstract knowledge of the schools. 'Through practice' writes Bernard Palissy (quoted from P. Rossi, *Philosophy, Technology and the Arts in the Early Modern Era*, New York 1970, p. 2 – the book contains many similar quotations and a thorough analysis of the situation from which they arose) 'I prove that the theories of many philosophers, even the most ancient and famous ones, are erroneous in many points.' Through practice Paracelsus showed that the medical knowledge of herbalists, country doctors, witches was superior to the knowledge of the scientific medicine of the time. Through practice navigators disproved the cosmological and climatological notions of the schools. It is interesting to see that the situation has not much changed. 'Through practice' acupuncturists and herbalists show that they can diagnose and heal illnesses whose effects scientific medicine recognizes but which it neither understands nor heals. 'Through practice' Thor Heyerdahl refuted scientific opinions about possibilities of navigation and seaworthiness of ships (cf. *The Ra Expeditions*, New York 1972, pp. 120, 155, 156, 122, 175, 261, 307 etc. concerning papyrus boats). 'Through practice' media produced effects which did not fit into the scientific world view and were ridiculed until a few fearless scientists proceded to examine them and proved their reality. [Even staid scientific organizations such as the American Association for the Advancement of Science now take them seriously and give them institutional recognition (incorporation of organizations dedicated to the study of parapsychological phenomena).] The rise of modern science has not eliminated the tension between extrascientific practice and school opinion, it has only given it a different content. School opinion is no longer Aristotle, it is not even restricted to a specific author, it is a body of doctrines, methods and experimental procedures that claims to possess the only reliable method for finding truth – and is constantly proven wrong in this claim (though the screening procedures mentioned in the text above make it difficult to discover major failures).

[30] For this and related fields cf. R. R. Hodson, ed., *The Place of Astronomy in the Ancient World*, Oxford 1974.

[31] For a survey cf. E. Mitchell, *op. cit.*

[32] Cf. the material in Chapters 1 and 2 of Lévi-Strauss, *The Savage Mind*. Physicians working with tribal healers have often admired their comprehension, knowledge and their quick understanding of new methods of healing (X-rays, for example).

to an 'objective' laboratory approach.[33] Nor is it surprising to find that ancient man had views worth considering. Stone Age man was already the fully developed *homo sapiens*, he was faced by tremendous problems which he solved with great ingenuity. Science is always praised because of its achievements. So let us not forget that the inventors of myth invented fire, and the means of keeping it. They domesticated animals, bred new types of plants, kept types separate to an extent that exceeds what is possible in today's scientific agriculture.[34] They invented rotation of fields and developed an art that can compare with the best creations of Western man. Not being hampered by specialization they found large scale connections between man and man and man and nature and relied on them to improve their science and their societies: the best ecological philosophy is found in the Stone Age. They crossed the oceans in vessels that were more seaworthy than modern vessels of comparable size and demonstrated a knowledge of navigation and the properties of materials that conflicts with scientific ideas but is, on trial, found to be correct.[35] They were aware of the role of change and their fundamental laws took this into account. It is only quite recently that science has returned to the Stone Age view of change after a long and dogmatic insistence on 'eternal laws of nature' that started with the 'rationalism' of the Presocratics and culminated towards the end of the last century. Moreover, these were not instinctive discoveries, they were the result of thought and speculation. 'There is abundant data which suggests not only that hunter–gatherers have adequate supplies of food but also that they enjoy quantities of leisure time, much more in fact than do modern industrial or farm workers, or even professors of archaeology.' There was abundant opportunity for 'pure thought'.[36] It is no good insisting that the discoveries of Stone Age man were due to an instinctive use of the correct scientific method. If they were, and if they led to correct results, then why did later scientists so often come to different conclusions? And, besides, there is no 'scientific method', as we have seen. Thus if science is praised because of its achievements, then myth must be praised a hundred times more fervently because *its* achievements were incomparably greater. The inventors of myth *started* culture while rationalists and scientists just

[33] Cf. Chapter 4 of *AM*.

[34] E. Anderson, *Plants, Man and Life*, London 1954.

[35] Cf. *Kon Tiki* and *The Ra Expeditions* by Thor Heyerdahl, esp. pp. 120, 122, 153, 132, 175, 206, 218f., 259 of the latter on the seaworthiness of papyrus and the proper construction of rafts.

[36] L. R. Binford and S. R. Binford, *New Perspectives in Archaeology*, Chicago 1968, p. 328. Cf. also the work of Marshall Sahlins.

changed it, and not always for the better.[37]

Assumption (b) can be refuted with equal ease: there is not a single important scientific idea that was not stolen from elsewhere. The Copernican Revolution is an excellent example. Where did Copernicus get his ideas? From ancient authorities, as he says himself. Who are the authorities that played a role in his thinking? Philolaos, among others, and Philolaos was a muddleheaded Pythagorean. How did Copernicus proceed when trying to make the ideas of Philolaos part of the astronomy of his time? By violating reasonable methodological rules. 'There is no limit to my astonishment' writes Galileo[38] 'when I reflect that Aristarchus and Copernicus were able to make reason so conquer sense that, in defiance of the latter, the former became mistress of their belief.' 'Sense', here, refers to the experience which Aristotle and others had used to show that the earth must be at rest. The 'reason' which Copernicus opposes to such arguments is the very mystical reason of Philolaos (and of the Hermeticists) combined with an equally mystical faith in the fundamental character of circular motion. Modern astronomy and modern dynamics could not have advanced without this unscientific use of antediluvian ideas.

While astronomy profited from Pythagoreanism and from the Platonic love for circles, medicine profited from herbalism, from the psychology, the metaphysics, the physiology of witches, midwives, cunning men, wandering druggists. It is well known that 16th and 17th century medical science, while theoretically hypertrophic, was quite helpless in the face of disease (and stayed that way for quite some time after the 'scientific revolution'). Innovators like Paracelsus fell back on earlier ideas and improved medicine. Everywhere science is enriched by unscientific methods and unscientific results while procedures which have often been regarded as essential parts of science are quietly suspended or circumvented.

---

[37] In Hesiod, who preserved earlier stages of thought, laws *come into existence* (rule of Zeus) and are the result of a *balance of opposing forces* (titans in fetters). They are the result of a dynamic equilibrium. In the 19th century laws were regarded as eternal and absolute, i.e. not due to a balance of mutually restricting entities. Hesiod's cosmology is far ahead of 19th century science.

[38] *Dialogue Concerning the Two Chief World Systems*, tr. Drake, Berkeley and Los Angeles 1954, p. 328. For details cf. the chapters on Galileo in *AM*.

10. Science is one Ideology among many and should be separated from the State just as Religion is now separated from the State

I started by stipulating that a free society is a society in which all traditions have equal rights and equal access to the centres of power.

This led to the objection that equal rights can be guaranteed only if the basic structure of society is 'objective', not influenced by undue pressures from any one of the traditions. Hence, rationalism will be more important than other traditions.

Now if rationalism and the accompanying views are not yet in existence or have no power then they cannot influence society as planned. Yet life is not chaos under such circumstances. There are wars, there is power-play, there are open debates between different cultures. The tradition of objectivity may therefore be introduced in a variety of ways. Assume it is introduced by an open debate – then, why should we change the form of debate at this point? Intellectuals say because of the 'objectivity' of their procedure – a pitiful lack of perspective, as we have seen. There is no reason to stick to reason even if it was reached by an open debate. There is even less reason to stick to it if it was imposed by force. This removes the objection.

The second objection is that though traditions may perhaps claim equal *rights* they do not produce equal *results*. This may be discovered by an open debate. The implication is that the excellence of science was established long ago – so why the fuss?

There are two replies to this objection. First that the comparative excellence of science has been anything but established. There are of course many *rumours* to that effect, but the *arguments* that are offered dissolve on closer inspection. Science does not excel because of its method for there is no method; and it does not excel because of its results: we know what science *does*, we have not the faintest idea whether other traditions could not do *much better*. So, we must find out.

To find out we must let all traditions freely develop side by side as is at any rate required by the basic stipulation of a free society. It is quite possible that an open debate about this development will find that some traditions have less to offer than others. This does not mean that they will be abolished – they will survive and keep their rights as long as there are people interested in them – it only means that for the time being their (material, intellectual, emotional etc.) products play a relatively small role. But what pleases once does not please always; and what aids traditions in one period does not aid them in others. The open debate and with it the examination of the favoured traditions will therefore continue: society is never identified with one particular tradition, and state and traditions are always kept separate.

The separation of state and science (rationalism) which is an essential part of this general separation of state and traditions cannot be introduced by a single political act and it should not be introduced in this way: many people have not yet reached the maturity necessary for living in a free society (this applies especially to scientists and other rationalists). People in a free society must decide about very basic issues, they must know how to assemble the necessary information, they must understand the purpose of traditions different from their own and the roles they play in the lives of their members. The maturity I am speaking about is not an intellectual virtue, it is a sensitivity that can only be acquired by frequent contacts with different points of view. It can't be taught in schools and it is vain to expect that 'social studies' will create the wisdom we need. But it can be acquired by participating in citizens' initiatives. This is why the *slow* progress, the *slow* erosion of the authority of science and of other pushy institutions that is produced by these initiatives is to be preferred to more radical measures: citizen initiatives are the best and only school for free citizens we now have.

11. Origin of the Ideas of this Essay

The problem of knowledge and education in a free society first struck me during my tenure of a state fellowship at the Weimar Institut zur Methodologischen Erneuerung des Deutschen Theaters (1946) which was a continuation of the Deutsches Theater Moskau under the directorship of Maxim Vallentin. Staff and Students of the Institut periodically visited theatres in Eastern Germany. A special train brought us from city to city. We arrived, dined, talked to the actors, watched two or three plays. After each performance the public was asked to remain seated while we started a discussion of what we had just seen. There were classical plays, but there were also new plays which tried to analyse recent events. Most of the time they dealt with the work of the resistance in Nazi Germany. They were indistinguishable from earlier Nazi plays eulogizing the activity of the Nazi underground in democratic countries. In both cases there were ideological speeches, outbursts of sincerity and dangerous situations in the cops and robbers tradition. This puzzled me and I commented on it in the debates: how is a play to be structured so that one recognizes it as presenting the 'good side'? What has to be added to the action to make the struggle of the resistance fighter appear morally superior to the struggle of an illegal Nazi in Austria before 1938? It is not sufficient to give him the 'right slogans' for then we take his superiority for granted, we do not show wherein it consists. Nor can his nobility, his 'humanity' be the distinguishing mark; every movement has scoundrels as well as noble people among its followers. A playwright may of course decide that sophistication is luxury in moral battles and give a black-white account. He may lead his followers to victory but at the expense of turning them into barbarians. What, then, is the solution? At the time I opted for Eisenstein and ruthless propaganda for the 'right cause'. I don't know whether this was because of any deep conviction of mine, or because I was carried along by events, or because of the magnificent art of Eisenstein. Today I would say that the choice must be left to the audience. The playwright presents characters and tells a story. If he errs it should be on the side of sympathy for his scoundrels, for circumstances and suffering play as large a role in the creation of evil and evil intentions as do those intentions themselves, and the general tendency is to emphasize the latter. The playwright (and his colleague, the teacher) must not try to anticipate the decision of the audience (of the pupils) or replace it by a decision of his own if they should turn out to be incapable of making up their own minds. *Under no circumstances must he try to be a 'moral force'.* A moral force, whether for good or for evil, turns people into slaves and slavery, even slavery in the service of The Good, or of God Himself is the most abject condition of all. This is how I see the situation today. However, it took me a long time before I arrived at this view.

After a year in Weimar I wanted to add the sciences and the humanities to the arts, and the theatre. I left Weimar and became a student (history, auxiliary sciences) at the famous *Institut für Osterreichische Geschichtforschung* which is part of the University of Vienna. Later on I added physics and astronomy and so finally returned to the subject I had decided to pursue before the interruptions of World War Two.

There were the following 'influences'.

(1) The *Kraft Circle*. Many of us science and engineering students were interested in the foundations of science and in broader philosophical problems. We visited philosophy lectures. The lectures bored us and we were soon thrown out because we asked questions and made sarcastic remarks. I still remember Professor Heintel advising me with raised arms: 'Herr Feyerabend, entweder sie halten das Maul, oder sie verlassen den

Vorlesungssaal!' We did not give up and founded a philosophy club of our own. Victor Kraft, one of my teachers, became our chairman. The members of the club were mostly students,[39] but there were also visits by faculty members and foreign dignitaries. Juhos, Heintel, Hollitscher, von Wright, Anscombe, Wittgenstein came to our meetings and debated with us. Wittgenstein who took a long time to make up his mind and then appeared over an hour late gave a spirited performance and seemed to prefer our disrespectful attitude to the fawning admiration he encountered elsewhere. Our discussions started in 1949 and proceeded with interruptions up to 1952 (or 53). Almost the whole of my thesis was presented and analysed at the meetings and some of my early papers are a direct outcome of these debates.

(2) The Kraft Circle was part of an organization called the *Austrian College Society*. The Society had been founded in 1945 by Austrian resistance fighters[40] to provide a forum for the exchange of scholars and ideas and so to prepare the political unification of Europe. There were seminars, like the Kraft Circle, during the academic year and international meetings during the summer. The meetings took place (and still take place) in Alpbach, a small mountain village in Tirol. Here I met outstanding scholars, artists, politicians and I owe my academic career to the friendly help of some of them. I also began suspecting that what counts in a public debate are not arguments but certain ways of presenting one's case. To test the suspicion I intervened in the debates defending absurd views with great assurance. I was consumed by fear – after all, I was just a student surrounded by bigshots – but having once attended an acting school I proved the case to my own satisfaction. The difficulties of *scientific* rationality were made very clear by

(3) *Felix Ehrenhaft* who arrived in Vienna in 1947. We, the students of physics, mathematics, astronomy had heard a lot about him. We knew that he was an excellent experimenter and that his lectures were performances on a grand scale which his assistants had to prepare for hours in advance. We knew that he had taught theoretical physics which was as exceptional for an experimentalist then as it is now. We were also familiar

---

[39] Many of them have now become scientists or engineers. Johnny Sogon is Professor of Mathematics at the University of Illinois, Henrich Eichorn (who also signed the anti-astrological encyclical mentioned above) director of New Haven observatory, Goldberger – de Buda adviser to electronics firms while Erich Jantsch who met members of our circle at the astronomical observatory has become a guru of dissident or pseudo-dissident scientists, trying to use old traditions for new purposes.

[40] Otto Molden, brother of Fritz Molden of the Molden publishing house, was for many years the dynamic leader and organizer.

with the persistent rumours that denounced him as a charlatan. Regarding ourselves as defenders of the purity of physics we looked forward to exposing him in public. At any rate our curiosity was aroused – and we were not disappointed.

Ehrenhaft was a mountain of a man, full of vitality and unusual ideas. His lectures compared favourably (or unfavourably, depending on the point of view) with the more refined performances of his colleagues. 'Are you dumb? Are you stupid? Do you really agree with everything I say?' he shouted at us who had intended to expose him but sat in silent astonishment at his performance. The question was more than justified for there were large chunks to swallow. Relativity and quantum theory were rejected at once, and almost as a matter of course for being idle speculation. In this respect Ehrenhaft's attitude was very close to that of Stark and Lenard both of whom he mentioned more than once with approval. But he went further than they and criticized the foundations of classical physics as well. The first thing to be removed was the law of inertia: undisturbed objects instead of going in a straight line were supposed to move in a helix. Then came a sustained attack on the principles of electromagnetic theory and especially on the equation div B = o. Then new and surprising properties of light were demonstrated – and so on and so forth. Each demonstration was accompanied by a few gently ironical remarks on 'school physics' and the 'theoreticians' who built castles in the air without considering the experiments which Ehrenhaft devised and continued devising in all fields and which produced a plethora of inexplicable results.

We had soon an opportunity to witness the attitude of orthodox physicists. In 1949 Ehrenhaft came to Alpbach. In that year Popper conducted a seminar on philosophy, Rosenfeld and M. H. L. Pryce taught physics and philosophy of physics (mainly from Bohr's comments on Einstein which had then just appeared), Max Hartmann biology, Duncan Sandys talked on problems of British politics, Hayek on economics and so on. There was Hans Thirring, the senior theoretical physicist from Vienna who constantly tried to impress on us that there were more important things than science and who had taught theoretical physics to Feigl, Popper as well as the present author. His son Walter Thirring, now Professor of Theoretical Physics in Vienna was also present – a very distinguished audience and a very critical one.

Ehrenhaft came well prepared. He set up a few of his simple experiments in one of the country houses of Alpbach and invited everyone he could lay hands on to have a look. Every day from two to three in the afternoon participants went by in an attitude of wonder and left the building (if they were theoretical physicists, that is) as if they had seen something obscene. Apart from these physical preparations Ehrenhaft also carried out, as was his habit, a beautiful piece of advertising. The day before his lecture he attended a fairly technical talk by von Hayek on 'The Sensory Order' (now available, in expanded form, as a book). During the discussion he rose, bewilderment and respect in his face, and started in a most innocent voice: 'Dear Professor Hayek. This was a marvellous, an admirable, a most learned lecture. I did not understand a single word . . .'. Next day his lecture had an overflow audience.

In this lecture Ehrenhaft gave a brief account of his discoveries adding general observations on the state of physics. 'Now gentlemen' he concluded triumphantly, turning to Rosenfeld and Pryce who sat in the front row – 'what can you say?' And he answered immediately. 'There is nothing at all you can say with all your fine theories. Sitzen muessen sie bleiben! Still muessen sie sein!'

The discussion, as was to be expected, was quite turbulent and it was continued for days with Thirring and Popper taking Ehrenhaft's side against Rosenfeld and Pryce. Confronted with the experiments the latter occasionally acted almost as some of Galileo's opponents must have acted when confronted with the telescope. They pointed out that no conclusions could be drawn from complex phenomena and that a detailed analysis was needed. In short, the phenomena were a *Dreckeffect* – a word that was heard quite frequently in the arguments. What was our attitude in the face of all this commotion?

None of us was prepared to give up theory or to deny its excellence. We founded a Club for the Salvation of Theoretical Physics and started discussing simple experiments. It turned out that the relation between theory and experiment was much more complex than is shown in textbooks and even in research papers. There are a few paradigmatic cases where the theory can be applied without major adjustments but the rest must be dealt with by occasionally rather doubtful approximations and auxiliary assumptions.[41] I find it quite interesting to remember how little effect all this had on us at the time. We continued to prefer abstractions as if the difficulties we had found had not been an expression of the nature of things but could be removed by some ingenious device, yet to be discovered. Only much later did Ehrenhaft's lesson sink in and our attitude at the time as well as the attitude of the entire profession provided me then with an excellent illustration of the nature of scientific rationality.

[41] *AM*, p. 63 on ad hoc approximations.

(4) *Philipp Frank* came to Alpbach a few years after Ehrenhaft. He undermined common ideas of rationality in a different way by showing that the arguments against Copernicus had been perfectly sound and in agreement with experience while Galileo's procedures were 'unscientific' when viewed from a modern standpoint. His observations fascinated me and I examined the matter further. Chapters 8 to 11 of *AM* are a late result of this study (I am a slow worker). Frank's work has been treated quite unfairly by philosophers like Putnam who prefer simplistic models to the analysis of complex historical events. Also his ideas are now commonplace. But it was he who announced them when almost everyone thought differently.

(5) In Vienna I became acquainted with some of the foremost Marxist intellectuals. This was the result of an ingenious PR job by Marxist students. They turned up – as did we – at all major discussions whether the subject was science, religion, politics, the theatre, or free love. They talked to those of us who used science to ridicule the rest – which was then my favourite occupation – and invited us to discussions of their own and introduced us to Marxist thinkers from all fields. I came to know Berthold Viertel, the director of the Burgtheater, Hanns Eisler, the composer and music theoretician and *Walter Hollitscher* who became a teacher and, later on, one of my best friends. When starting to discuss with Hollitscher I was a raving positivist, I favoured strict rules of research and had only a pitying smile for the three basic principles of dialectics which I read in Stalin's little pamphlet on dialectical and historical materialism. I was interested in the realist position, I had tried to read every book on realism I could lay hands on (including Külpe's excellent *Realisierung* and, of course, *Materialism and Empiriocriticism*) but I found that the arguments for realism worked only when the realist assumption had already been introduced. Külpe, for example, emphasized the distinction between impression and the thing the impression is about. The distinction gives us realism only if it characterizes real features of the world – which is the point at issue. Nor was I convinced by the remark that science is an essentially realistic enterprise. Why should science be chosen as an authority? And were there not positivistic interpretations of science? The so-called 'paradoxes' of positivism, however, which Lenin exposed with such consummate skill did not impress me at all. They arose only if the positivist and the realist mode of speech were mixed and exposed their difference. They did not show that realism was better though the fact that realism came with common speech gave the impression that it was.

Hollitscher never presented an argument that would lead, step by step, from positivism into realism and he would have regarded the attempt to produce such an argument as philosophical folly. He rather developed the realist position itself, illustrated it by examples from science and commonsense, showed how closely it was connected with scientific research and everyday action and so revealed its strength. It was of course always possible to turn a realistic procedure into a positivistic procedure by a judicious use of ad hoc hypotheses and ad hoc meaning changes and I did this frequently, and without shame (in the Kraft Circle we had developed such evasions into a fine art). Hollitscher did not raise semantic points, or points of method, as a critical rationalist might have done, he continued to discuss concrete cases until I felt rather foolish with my abstract objections. For I saw now how closely realism was connected with facts, procedures, principles I valued and that it *had helped to bring them about* while positivism merely *described* the results in a rather complicated way after they had been found: realism had fruits, positivism had none. This at least is how I would speak today, long *after* my realist conversion. At the time I became a realist not because I was convinced by any particular argument, but because the sum total of realism plus the arguments in favour of it plus the ease with which it could be applied to science and many other things I vaguely felt but could not lay a finger on[42] finally *looked better to me* than the sum total of positivism plus the arguments one could offer for it plus ... etc. etc. The comparison and the final decision had much in common with the comparison of life in different countries (weather, character of people, melodiousness of language, food, laws, institutions, weather etc. etc.) and the final decision to take a job and to start life in one of them. Experiences such as these have played a decisive role in my attitude towards rationalism.

While I accepted realism I did not accept dialectics and historical materialism – my predilection for abstract arguments (another positivist hangover) was still too strong for that. Today Stalin's rules seem to me preferable by far to the complicated and epicycle-ridden standards of our modern friends of reason.

[42] I remember that Reichenbach's answer to Dingler's account of relativity played an important part: Dingler extrapolated from what could be achieved by simple mechanical operations (manufacture of a Euclidian plain surface, for example) while Reichenbach pointed out how the actual structure of the world would modify the results of these operations in the large. It is of course true that Reichenbach's account can be interpreted as a more efficient predictive machine and that it seemed impressive to me only because I did not slide into such an interpretation. Which shows to what extent the force of arguments depends on irrational changes of attitude.

From the very beginning of our discussion Hollitscher made it clear that he was a communist and that he would try to convince me of the intellectual and social advantages of dialectical and historical materialism. There was none of the mealy-mouthed 'I may be wrong, you may be right – but together we shall find the truth' talk with which 'critical' rationalists embroider their attempts at indoctrination but which they forget the moment their position is seriously endangered. Nor did Hollitscher use unfair emotional or intellectual pressures. Of course, he criticized my attitude and he still does but our personal relations have not suffered from my reluctance to follow him in every respect. This is why Walter Hollitscher is a teacher while Popper whom I also came to know quite well is a mere propagandist.

At some point of our acquaintance Hollitscher asked me whether I would like to become a production assistant of Brecht – apparently there was a position available and I was being considered for it. I declined. This, I think, was one of the biggest mistakes of my life. Enriching and changing knowledge, emotions, attitudes through the arts now seems to me a much more fruitful enterprise and also much more humane than the attempt to influence minds (and nothing else) by words (and nothing else). If today only about 10% of my talents are developed then this is due to a wrong decision at the age of 25.

(6) During a lecture (on Descartes) I gave at the Austrian College Society I met *Elizabeth Anscombe*, a powerful and, to some people, forbidding British philosopher who had come to Vienna to learn German for her translation of Wittgenstein's works. She gave me manuscripts of Wittgenstein's later writings and discussed them with me. The discussions extended over months and occasionally proceeded from morning over lunch until late into the evening. They had a profound influence upon me though it is not at all easy to specify particulars. On one occasion which I remember vividly Anscombe, by a series of skilful questions, made me see how our conception (and even our perceptions) of well-defined and apparently self-contained facts may depend on circumstances not apparent in them. There are entities such as physical objects which obey a 'conservation principle' in the sense that they retain their identity through a variety of manifestations and even when they are not present at all while other entities such as pains and after images are 'annihilated' with their disappearance. The conservation principles may change from one developmental stage of the human organism to another[43] and they

---

[43] Cf. *AM*, pp. 227ff.

may be different for different languages (cf. Whorff's 'covert classifications' as described in Chapter 17 of *AM*). I conjectured that such principles would play an important role in science, that they might change during revolutions and that deductive relations between pre-revolutionary and post-revolutionary theories might be broken off as a result. I explained this early version of incommensurability in Popper's seminar (1952) and to a small group of people in Anscombe's flat in Oxford (also in 1952 with Geach, von Wright and L. L. Hart present) but I was not able to arouse much enthusiasm on either occasion.[44] Wittgenstein's emphasis on the need for concrete research and his objections to abstract reasoning ('Look, don't think!') somewhat clashed with my own inclinations and the papers in which his influence is noticeable are therefore mixtures of concrete examples and sweeping principles.[45] Wittgenstein was prepared to take me on as a student in Cambridge but he died before I arrived in England. Popper became my supervisor instead.

(7) I had met *Popper* in Alpbach in 1948. I admired his freedom of manners, his cheek, his disrespectful attitude towards the German philosophers who gave the proceedings weight in more senses than one, his sense of humour (yes, the relatively unknown Karl Popper of 1948 was very different from the established Sir Karl of later years) and I also admired his ability to restate ponderous problems in simple and journalistic language. Here was a free mind, joyfully putting forth his ideas, unconcerned about the reaction of the 'professionals'. Things were different as regards these ideas themselves. The members of our Circle knew deductivism from Kraft who had developed it before Popper,[46] the falsificationist philosophy was taken for granted in the physics seminar of the conference under the chairmanship of Arthur March and so we did not understand what all the fuss was about. 'Philosophy must be in a desperate state' we said 'if trivialities such as these can count as major discoveries'. Popper himself did not seem to think too much of his philosophy of science at the time for when asked to send us a list of publications he included the *Open Society* but not the *Logic of Scientific Discovery*.

While in London I read Wittgenstein's *Philosophical Investigations* in

---

[44] For details cf. Part One, Section 7 of this volume.

[45] For details cf. my comments on these papers in *Der Wissenschaftstheoretische Realismus und die Autorität der Wissenschaften*, Vieweg Wiesbaden 1978.

[46] Cf. my review of Kraft's *Erkenntnislehre* in *BJPS*, Vol. 13 (1963), pp. 319ff. and esp. p. 321, second paragraph. Cf. also the references in Popper, *Logic of Scientific Discovery*.

detail. Being of a rather pedantic turn of mind I rewrote the book so that it looked more like a treatise with a continuous argument. Part of this treatise was translated by Anscombe into English and published as a review by *Phil. Rev.* in 1955. I also visited Popper's seminar at the LSE. Popper's ideas were similar to those of Wittgenstein but they were more abstract and anaemic. This did not deter me but increased my own tendencies to abstraction and dogmatism. At the end of my stay in London Popper invited me to become his assistant. I declined despite the fact that I was broke and did not know where my next meal was going to come from. My decision was not based on any clearly recognizable train of thought but I guess that having no fixed philosophy I preferred stumbling around in the world of ideas at my own speed to being guided by the ritual of a 'rational debate'. Two years later Popper, Schrödinger and my own big mouth got me a job in Bristol where I started lecturing on the philosophy of science.

(8) I had studied theatre, history, mathematics, physics and astronomy, I had never studied philosophy. The prospect of having to address a large audience of eager young people did not exactly fill my heart with joy. One week before the lectures started I sat down and wrote everything I knew on a piece of paper. It hardly filled a page. Agassi came up with some excellent advice: 'Look Paul' he said 'the first line, this is your first lecture; the second line, this is your second lecture – and so on.' I took his advice and fared rather well except that my lectures became a stale collection of wisecracks by Wittgenstein, Bohr, Popper, Dingler, Eddington and others. While in Bristol I continued my studies of the quantum theory. I found that important physical principles rested on methodological assumptions that are violated whenever physics advances: physics gets authority from ideas it propagates but never obeys in actual research, methodologists play the role of publicity agents whom physicists hire to praise their results but whom they would not permit access to the enterprise itself. That falsificationism is not a solution became very clear in discussions with David Bohm who gave a Hegelian account of the relation between theories, their evidence, and their successors.[47] The material of Chapter 3 of *AM* is the result of these discussions (I first published it in 1961).[48] Kuhn's remarks on the omnipresence of anomalies fitted these

---

[47] I have explained the Hegelianism of Bohm in the essay 'Against Method' which appeared in Vol. iv of the *Minnesota Studies for the Philosophy of Science* (1970).

[48] Popper once remarked (in a discussion at the Minnesota Center for the Philosophy of Science in the year 1962) that the example of Brownian motion is just another version of Duhem's example (conflict between specific laws such as Kepler's laws and general theories such as Newton's theory). But there is a most important difference. The deviations from

difficulties rather nicely[49] but I still tried to find general rules that would cover all cases[50] and non-scientific developments as well.[51] Two events made me realize the futility of such attempts. One was a discussion with Professor C. F. von Weizsäcker in Hamburg (1965) on the foundations of the quantum theory. Von Weizsäcker showed how quantum mechanics arose from concrete research while I complained, on general methodological grounds, that important alternatives had been omitted. The arguments supporting my complaint were quite good – they are the arguments summarized in Chapter 3 of *AM* – but it was suddenly clear to me that imposed without regard to circumstances they were a hindrance rather than a help: a person trying to solve a problem whether in science or elsewhere *must be given complete freedom* and cannot be restricted by any demands, norms, however plausible they may seem to the logician or the philosopher who has thought them out in the privacy of his study. Norms and demands must be checked by research, not by appeal to theories of rationality. In a lengthy article[52] I explained how Bohr had used this philosophy and how it differs from more abstract procedures. Thus Professor von Weizsäcker has prime responsibility for my change to 'anarchism' – though he was not at all pleased when I told him so in 1977.

(9) The second event that prompted me to move away from rationalism and to become suspicious of all intellectuals was quite different. To explain it, let me start with some more general observations. The way in which social problems, problems of energy distribution, ecology, education, care for the old and so on are 'solved' in our societies can be roughly described in the following way. A problem arises. Nothing is done about it. People get concerned. Politicians broadcast this concern. Experts are called in. They develop a plan or a variety of plans. Power-groups with experts of their own effect various modifications until a watered down version is accepted and realized. The role of experts in this process has gradually increased. Intellectuals have developed theories about the application of science to social problems. 'To get ideas' they ask other

---

Kepler's laws are in principle observable ('in principle' here meaning 'given the known laws of nature') while the microscopic deviations from the second law of thermodynamics are not (measuring instruments are subjected to the same fluctuations as the things they are supposed to measure). Here we *cannot* do without an alternative theory.

[49] I read Kuhn's book in manuscript in 1960 and discussed it extensively with Kuhn.

[50] Cf. the account in 'Reply to Criticism', *Boston Studies*, Vol. ii, 1965.

[51] Cf. 'On the Improvement of the Sciences and the Arts and the Possible Identity of the Two' in *Boston Studies*, Vol. iii, 1967.

[52] 'On a Recent Critique of Complementarity', *Philosophy of Science* 1968/69 (two parts).

intellectuals, or politicians. Only rarely does it occur to them that it is not their business *but the business of those immediately concerned* to decide the matter. They simply take it for granted that their ideas and those of their colleagues are the only important ones and that people have to adapt to them. What has this situation got to do with me?

From 1958 on I was a Professor of Philosophy at the University of California in Berkeley. My function was to carry out the educational policies of the State of California which means I had to teach people what a small group of white intellectuals had decided was knowledge. I hardly ever thought about this function and I would not have taken it very seriously had I been informed. I told the students what I had learned, I arranged the material in a way that seemed plausible and interesting to me – and that was all I did. Of course, I had also some 'ideas of my own' – but these ideas moved in a fairly narrow domain (though some of my friends said even then that I was going batty).

In the years 1964ff. Mexicans, Blacks, Indians entered the university as a result of new educational policies. There they sat, partly curious, partly disdainful, partly simply confused hoping to get an 'education'. What an opportunity for a prophet in search of a following! What an opportunity, my rationalist friends told me, to contribute to the spreading of reason and the improvement of mankind! What a marvellous opportunity for a new wave of enlightenment! I felt very differently. For it dawned on me that the intricate arguments and the wonderful stories I had so far told to my more or less sophisticated audience might just be dreams, reflections of the conceit of a small group who had succeeded in enslaving everyone else with their ideas. Who was I to tell these people what and how to think? I did not know their problems though I knew they had many. I was not familiar with their interests, their feelings, their fears though I knew that they were eager to learn. Were the arid sophistications which philosophers had managed to accumulate over the ages and which liberals had surrounded with schmaltzy phrases to make them palatable the right thing to offer to people who had been robbed of their land, their culture, their dignity and who were now supposed to absorb patiently and then to repeat the anaemic ideas of the mouthpieces of their oh so human captors? They wanted to know, they wanted to learn, they wanted to understand the strange world around them – did they not deserve better nourishment? Their ancestors had developed cultures of their own, colourful languages, harmonious views of the relation between man and man and man and nature whose remnants are a living criticism of the tendencies of separation, analysis, self-centredness inherent in

Western thought. These cultures have important achievements in what is today called sociology, psychology, medicine, they express ideals of life and possibilities of human existence. Yet *they were never examined with the respect they deserved* except by a small number of outsiders, they were ridiculed and replaced as a matter of course first by the religion of brotherly love and then by the religion of science or else they were defused by a variety of 'interpretations' (cf. Section 2 above). Now there was much talk of liberation, of racial equality – but what did it mean? Did it mean the equality of these traditions and the traditions of the white man? It did not. Equality meant that the members of different races and cultures now had the wonderful chance to participate in the white man's manias, they had the chance to participate in his science, his technology, his medicine, his politics. These were the thoughts that went through my head as I looked at my audience and they made me recoil in revulsion and terror from the task I was supposed to perform. For the task – this now became clear to me – was that of a very refined, very sophisticated slave-driver. And a slavedriver I did not want to be.

Experiences such as these convinced me that intellectual procedures which approach a problem through concepts and abstract from every-thing else are on the wrong track and I became interested in the reasons for the tremendous power this error has now over minds. I started examining the rise of intellectualism in Ancient Greece and the causes that brought it about. I wanted to know what it is that makes people who have a rich and complex culture fall for dry abstractions and mutilate their traditions, their thought, their language so that they can accommo-date the abstractions. I wanted to know how intellectuals manage to get away with murder – for it is murder, murder of minds and cultures that is committed year in year out at schools, universities, educational missions in foreign countries. The trend must be reversed, I thought, we must start learning from those we have enslaved for they have much to offer and at any rate, they have the right to live as they see fit even if they are not as pushy about their rights and their views as their Western Conquerors have always been. In 1964–5 when these ideas first occurred to me I tried to find an *intellectual* solution to my misgivings that is, I took it for granted that it was up to *me* and the likes of me to devise educational policies for other people. I envisaged a new kind of education that would live from a rich reservoir of different points of view permitting the choice of traditions most advantageous to the individual. The teacher's task would consist of facilitating the choice, not in replacing it by some 'truth' of his own. Such a reservoir, I thought, would have much in common with a *theatre* of ideas as imagined by Piscator and Brecht and it would lead to the development of a great variety of means of presentation. The 'objective' scientific account would be one way of presenting a case, a play another way (remember that for Aristotle tragedy is 'more philosophical' than history because it reveals the *structure* of the historical process and not only its accidental details) a novel still another way. Why should knowledge be shown in the garment of academic prose and reasoning? Had not Plato observed that written sentences in a book are but transitory stages of a complex process of growth that contains gestures, jokes, asides, emotions and had he not tried to catch this process by means of the dialogue? And were there not different forms of knowledge, some much more detailed and realistic than what arose as 'rationalism' in the 7th and 6th century in Greece? Then there was *Dadaism*. I had studied Dadaism after the Second World War. What attracted me to this movement was the style its inventors used when not engaged in Dadaistic activities. It was clear, luminous, simple without being banal, precise without being narrow; it was a style adapted to the expression of thought as well as of emotion. I connected this style with the Dadaistic exercises themselves. Assume you tear language apart, you live for days and weeks in a world of cacophonic sounds, jumbled words, nonsensical events. Then, after this preparation, you sit down and write: 'the cat is on the mat'. This simple sentence which we usually utter without thought, like talking machines (and much of our talk is indeed routine) now seems like the creation of an entire world: God said let there be light, and there was light. Nobody in modern times has understood the miracle of language and thought as well as the Dadaists for nobody has been able to imagine, let alone create a world in which they play no role. Having discovered the nature of a *living order*, of a reason that is not merely mechanical, the Dadaists soon noticed the deterioration of such an order into routine. They diagnosed the deterioration of language that preceded the First World War and created the mentality that made it possible. After the diagnosis their exercises assumed another, more sinister meaning. They revealed the frightening similarity between the language of the foremost commercial travellers in 'importance', the language of philosophers, politicians, theologians, and brute inarticulation. The praise of honour, patriotism, truth, rationality, honesty that fills our schools, pulpits, political meetings *imperceptibly merges into inarticulation* no matter how much it has been wrapped into literary language and no matter how hard its authors try to copy the style of the classics and the authors themselves are in the end hardly distinguishable

from a pack of grunting pigs. Is there a way to prevent such deterioration?
I thought there was. I thought that regarding all achievements as transitory, restricted *and personal* and every truth as *created* by our love for it
and not as 'found' would prevent the deterioration of once promising
fairy tales and I also thought that it was necessary to develop a new
philosophy or a new religion to give substance to this unsystematic
conjecture.

I now realize that these considerations are just another example of
intellectualistic conceit and folly. It is conceited to assume that one has
solutions for people whose lives one does not share and whose problems
one does not know. It is foolish to assume that such an exercise in distant
humanitarianism will have effects pleasing to the people concerned. From
the very beginning of Western Rationalism intellectuals have regarded
themselves as teachers, the world as a school and 'people' as obedient
pupils. In Plato this is very clear. The same phenomenon occurs among
Christians, Rationalists, Fascists, Marxists. Marxists no longer try to
learn from those they want to liberate, they attack each other about
interpretations, viewpoints, evidence and take it for granted that the
resulting intellectual hash will make fine food for the natives (Bakunin
was aware of the doctrinarian tendencies of contemporary Marxism and
he intended to return all power – power over ideas included – to the
people immediately concerned). My own view differed from those just
mentioned but it was still a VIEW, an abstract fancy I had invented and
now tried to sell without having shared even an ounce of the lives of the
receivers. This I now regard as insufferable conceit. So – what remains?

Two things remain. I could start *participating* in some tradition and
try to reform it from the inside. This, I think, is important. The time
when Great Minds associating with Great Powers of Society could run
the lives of the rest even in an ever so gentle way slowly comes to an end
(this excludes Germany). More and more civilizations enter the stage of
world politics, more and more traditions are regained by people living
inside Western Societies. A person can either participate in these traditions (if they will have him) or shut up – he can no longer address them
as if they were students in a classroom. For a long time now I have been
a somewhat erratic member of a pseudo-scientific tradition – so I could
try to encourage from within those tendencies I find sympathetic. This
would agree with my inclination to use the *history of ideas* to explain
puzzling phenomena and to experiment with forms of expression different from scholastic prose to present and/or expose ideas. I have not

much enthusiasm for such work especially as I think that fields such as the philosophy of science, or elementary particle physics, or ordinary language philosophy, or Kantianism should not be reformed, but should be allowed to die a natural death (they are too expensive and the money spent on them is needed more urgently elsewhere). Another possibility is to start a career as an *entertainer*. This is very attractive to me. Bringing a faint smile to the faces of people who have been hurt, disappointed, depressed, who are paralysed by some 'truth' or by the fear of death seems to me an achievement infinitely more important than the most sublime intellectual discovery: Nestroy, George S. Kaufman, Aristophanes, on my scale of values range far above Kant, Einstein and their anaemic imitators. These are the possibilities. What shall I do? Only time will tell. . . .

Part Three

Conversations with Illiterates

Chapter 1

Reply to Professor Agassi

with a footnote for Rom Harré and a postcript

Berkeley, July 15, 1975

Dear Joske,*

There are three things which never fail to amaze me when reading reviews of my book: the disregard for argument, the violence of the reaction, the general impression I seem to make on my readers, and especially on 'rationalists'.

As I see it, my book is a longwinded and rather pedestrian attempt to criticize certain ideas about science and rationality, to reveal the idols behind the ideas, and to put them in their proper place. Not being as blinded by slogans as my rationalist critics seem to be I *investigate*, and I report the results of my investigation. My investigation is far from comprehensive. The most important problem of the relation between reason and *faith* is not even touched upon. What I do is this. I compare three idols: Truth, Honesty, Knowledge (or Rationality) and their methodological ramifications with a fourth idol, Science, I find that they conflict, and I conclude that it is time to take a fresh look at all of them. At any rate, neither science, nor rationalism have now sufficient authority to exclude myth, or 'primitive' thought, or the cosmologies behind the various religious creeds. Any claim to such authority is illegal and must be rejected, if necessary, by political means. I would say that my book contains 85% exposition and argument, 10% conjecture, and 5% rhetoric. There are long passages devoted to the description of fact and procedure.

Now the strange thing is that hardly any review I have read deals with this material. The only passages the reviewers seem to perceive are places where, with a sigh of relief, I stop reasoning and engage in a little rhetoric.[1] This means either that rationalists do not recognize an argu-

---

* This reply and the corresponding review appeared in *Philosophia*, March 1976.

[1] An example is Professor Rossi's article 'Hermeticism and Rationality in the Scientific Revolution' published in Bonelli–O'Shea, eds. *Reason, Experiment and Mysticism in the Scientific Revolution*, New York 1975, pp. 247–73. Professor Rossi carries on an unhappy

ment when they see one, or that they regard rhetoric as more important than argument, or else that something in my book so jars their thought and confounds their perception that dreams and hallucinations replace the reality in front of them. Your article, my dear Joske, is a perfect example of what I mean. I am very grateful that you are so deeply concerned about my book and that you have put so much time, energy, and especially imagination into the review. But, alas, I hardly recognized myself in the terrible portrait that glared at me from its pages. For you, my book seems to be a mixture of *Die Räuber* and *Ubu Roi*, combining the 'hate blasts' of the first with the cheerful nonsense of the second. Of course, you are very good, you almost succeeded in convincing me that I was a 'super revolutionary, in politics as well as in methodology' – but the illusion did not last very long. A look at my book, and I saw that I was mistaken and that you were mistaken. How did this mistake arise? And, having realized it, how can I prevent you and my future readers from repeating it? How can I wake you up, make you open your eyes so that you see what I have written and do not at once wander off into a dream world of your own? I do not know the recipe, but I shall try. And I ask your and the readers' indulgence when in the attempt to make myself understood I shall often be longwinded and tediously repetitive.

According to you I am a 'super revolutionary, in politics as well as in methodology' and my 'ideal is totalitarian China'.

The first sentence of my book reads (page 17, text of the Introduction): 'The following essay is written in the conviction that *anarchism*, while perhaps not the most attractive *political* philosophy, is certainly excellent medicine for *epistemology* and for the *philosophy of science*.'

Now I admit – one does not always read first sentences very carefully; one passes them over, one wants to go on to the more important parts of the book and see what surprises the author has hidden there. I also admit that people less pedantic than I do not turn every sentence into a package

---

love affair with rationalism. There is much love, there is little understanding. The philosophers and historians whom Rossi criticizes have produced many and detailed arguments to support their point of view. Their papers contain these arguments, the results of the arguments, and a sometimes rather colourful summation of the results. Professor Rossi recognizes the colourful summaries but he does not seem to have the ability to recognize an argument. Moreover, he rejects the summaries not because he possesses arguments of his own, but because he does not like them, or does not *seem* to like them, for even in the domain of *liking* he is not quite sure which way to go. On page 266 he speaks of the 'Neoromantic Revolt against science' and clearly disapproves of it. But on page 247 he complains about the fact that my interpretation of Galileo has been received 'enthusiastically' '*even* in Italy' implying that a 20th century Italian is better equipped to understand the Spirit of Galileo than a 20th century Viennese – a typically Romantic idea.

stuffed with information, they give the reader some leeway and permit him to get alowly acquainted with their style. So, I should perhaps be grateful to you for reading me as if I were a better and more elegant writer than I actually am. But, alas, at the present moment my gratitude is almost overwhelmed by my wish to be understood, and so the reader must bear with me when I explain the sentence in somewhat greater detail.

What does it say?

It says that I regard anarchism as 'excellent medicine for epistemology and the philosophy of science'.

Note the careful qualification. I do not say that epistemology should become anarchic, or that the philosophy of science should become anarchic. I say that both disciplines should receive anarchism *as a medicine*. Epistemology is sick, it must be cured, and the medicine is anarchy. Now medicine is not something one takes all the time. One takes it for a certain period of time, *and then one stops*. To make sure that this is the way in which I shall be understood I repeat the restriction at the end of the Introduction. In the last but one sentence I say: 'There may, of course, come a time when it will be necessary to give reason a temporary advantage and when it will be wise to defend *its* rules to the exclusion of everything else'. And then I continue: 'I do not think that we are living in such a time today'. *Today* epistemology is sick and in need of a medicine. The medicine is anarchism. Anarchism, I say, will heal epistemology and *then* we may return to a more enlightened and more liberal form of rationality. So far the first qualification that is contained in the first sentence of my book.

There are two more qualifications.

I say that anarchism is 'perhaps not the most attractive *political* philosophy'. Qualification one: I intend to discuss the role of anarchism in epistemology and in the philosophy of science, I am not too enthusiastic about political anarchism. Qualification two: however, I may be mistaken in my lack of enthusiasm (anarchism, 'while *perhaps* not . . .'). So far the first sentence of my book. Does it not sound very different from the picture of a 'super revolutionary, in politics as well as in methodology'?

How does the difference arise? The answer is fairly simple. When reading my book you omit qualifications which I either imply, or state explicitly. These qualifications are quite important. They are the essence of what I want to say. I think that there is very little we can say 'in general', that the observations we make and the advice we must give take a specific (historical, social psychological, physical etc.) situation into account, that we cannot proceed unless we have studied this situation in detail. (This, incidentally, is the rationale behind the slogan 'anything goes': if you want advice that remains valid, no matter what, then the advice will have to be as empty and indefinite as 'anything goes'.) Any statement I make has this *specific* character, the qualifications being either implicit in the context, or contained in the statement itself. You show no such discretion. Moving nimbly from page to page you notice only the phrases which shock you and you overlook the qualifications and the arguments that might have alleviated the shock. Let me mention another example to illustrate this procedure of selective reading.

In Chapter 4 I look favourably upon a certain episode in the relation between Party and Experts in the Communist China of the Fifties and I advise the democratic bodies of today to act in a similar manner. I think that *on that occasion* the Party acted reasonably and I suggest that democracies combat the chauvinism of their own experts in exactly the same manner. In your review this limited and concrete suggestion becomes 'Feyerabend's ideal is totalitarian China'. ('Chairman Mao's terror was not rejected' you write a little later. Of course it was not rejected; and why not? Because this was not the topic of my argument.[2])

However – you don't limit yourself to oversights. You not only omit, you also add, and in a most imaginative manner.

In my book I quote Lenin as a person well acquainted with the complexities of what some people call 'methodology'. I call him an 'intelligent and thoughtful observer' and I add in a footnote that he 'can give useful

---

[2] Elsewhere you say: 'there is no reference to Nazism, Fascism, even to the Spanish Civil War, not to mention racism'. True. And I did not mention burlesque either. And why not? Because there was not enough room, and because the items have hardly anything to do with my main topic which is: *epistemological* anarchism. Of course, you believe that irrationalism and anarchism *are bound to* lead to all these things (even to Auschwitz, as you said in a talk in Germany – really, Joske!), you expect that an author who recommends anarchistic moves is aware of such dangers, and so you think he has an obligation to comment on them. This would be acceptable reasoning if one could be sure that rationalism is free from dangers of this kind. *But it is not.* Quite the contrary. *The unrelenting driving force of reason is much more likely to stay with an antihumanitarian idea once it has been conceived than the quickly moving procedure of the anarchist.* Robespierre was a rationalist, not an anarchist; the inquisitors of the 16th and 17th centuries who burned tens of thousands of victims were rationalists, not anarchists; Urbach, who has found some very sophisticated arguments in favour of a very refined racism is a rationalist, even a critical rationalist, not an anarchist; the trouble with the Spanish Civil War was not the presence of anarchists but the fact that despite the majority they possessed they refused to form a government and so left the stage to more 'rational' politicians; and don't forget that the word 'god *fearing*' arose in Greece only after Xenophanes had replaced the Homeric gods by a more rational account of Being which was a forerunner of Parmenides' monstrous One and this change of attitude towards God was a direct consequence of the increase of rationalism. If I had your talent for generalization I would say that

advice to everyone, philosophers of science included'. Let me pass over the fact that in your review this becomes 'Lenin is the greatest methodologist of them all' – for such a change of emphasis is still within the limits of poetic licence as practised by you and noted above. But you continue: 'He means Marcuse, of course, but he says Lenin'. I must confess, I was absolutely flabbergasted to read this remark. I mean Marcuse, and 'of course'? How on earth did Marcuse get into the discussion? Do I mention him anywhere in my book? I look up the index, yes, I do, in a footnote, on page 27. I turn to page 27, for I have already forgotten how and why I mention him. On page 27, I find, I quote from a rather good introduction to Hegel which Marcuse wrote some time ago – and that is all. Did I mention Marcuse in other writings? Yes, I did, in the *essay* 'Against Method' which precedes the book, but in a severely critical manner. Besides, why should I 'of course' mean a university professor and third rate intellectual when speaking about a first rate thinker, writer, and politician? Especially in view of the fact that I prefer people who are aware of the complex connections between different fields to those who are content with simpleminded models?[3]

There is an even more amusing example of your tendency to let your mind wander when reading a book, and it occurs in connection with an autobiographical remark of mine. I write (page 126, end of n. 19): 'I still remember my disappointment when, having built a reflector with an alleged linear magnification of about 150, I found the moon only about five times enlarged, and situated quite close to the ocular'. This was to illustrate the difference between the predictions of *geometrical* optics and what one actually *sees* when looking into a telescope. You write: 'in 1937

---

rationalists are much more likely to build an Auschwitz than anarchists who, after all, want to remove all kinds of repression, repression by reason included. It is not rationalism, it is not law and order which prevents cruelty, but 'the unreasoning impulse of human kindness' as George Lincoln Burr said in a letter to A. D. White who tried to explain the disappearance of the witchcraft mania by the rise of rationalism. 'But kindness is an *irrational force*.'

[3] The reader should not be misled, as you obviously are, by my frequent praise of leftist politicians. I praise them not because they are on the left but because they are thinkers *as well as* politicians, theoreticians *as well as* men of action, and because their experience with this world has made their philosophy realistic and flexible. It is not my fault that there are no comparable figures on the right, or in the centre and that intellectuals, with the sole exception of Hegel, have been content with admiring, or destroying each other's castles in the air. Considering my reasons for choice I might also have chosen great religious figures such as Church Fathers in my examples – and I did so in some earlier writings of mine where I praised St. Irenaeus, Tertullian (superb intellect!), St. Augustine, St. Athanasius, and others. Even a Bossuet is preferable by far to the professional scribblers of today who extol 'ideas' but have very little to say about the fears and the needs of soul and body.

(the date when I started my observations) Austria was no kind place for a lonely youth whose scientific escapades met, one might imagine, with little or no understanding for his disappointment – not even from high school teachers whose minds were elsewhere anyway. This possibility might explain a lot of the feeling shown in the present exposé of the view of science as pretty-pretty . . .'. It is very nice of you, dear Joske Agassi, to give such a moving account of my childhood and such a generous explanation of the 'hate blasts' you seem to encounter in my book. But you again give me credit where credit is not due. Far from 'meeting with little or no understanding', my 'scientific escapades' were *caused* by an excellent physics teacher in high school who inspired us all to build meridian instruments, sundials, telescopes and who made me an official observer for the Swiss Center for Solar Activity at the tender age of 14 (it was at his lecture course at the university that I gave my very first public lecture on my thirteenth birthday). Now remember: it is one thing to write a fictional account of somebody's life and ideas for entertainment, or in order to direct attention to achievements which would otherwise remain unnoticed. But it is quite a different thing to make such an account (Lenin meaning 'of course' Marcuse; Chinese totalitarianism as a political ideal; all this fostered by youthful astronomical frustrations) the basis of a review. *I* can afford such extravaganzas, at least the I you seem to perceive in my book can – but you, my dear Joske cannot, for you are a rationalist, and tied to more severe standards.

So much about your failings as a reader and reviewer. There is another item I want to discuss before proceeding to more substantial issues, and it is this.

Many readers, and you, too, are disturbed by my way of saying things. 'I think *what* you say is allright; but I think *how* you say it is wrong' writes our common friend Henryk Skolimowski in a letter I just received. You speak of my 'hate blasts' and of my 'scathing attacks'. You do not identify the former, but you give a page number for the latter, so I again turn to the book and read. And I am again amazed at the difference between your perception and mine. For the page to which you refer contains a very *mild* (though concisely formulated) criticism of Clavius, Grienberger, and father McMullin, another common friend of ours. We obviously look at things in a very different way.[4]

[4] Rom Harré (*Mind* 1977, p. 295) has a similar complaint. 'The asides about women, about friends and colleagues, indeed about everyone who is likely to place trust in someone else, exhibit an attitude of contempt . . . Professor Feyerabend claims a total licence for himself and other great souls such as Galileo, a licence which includes the right to make offensive and

The reason, I think, is that we have different ideas about *style*. *You* (and many other readers) are fond of a perhaps lively, vigorous, but still *scholarly* style. *I* find such a style with its neat innuendos and its civilized strangulation of the opponent too desiccated and also too dishonest (strange word for me to use – eh?) for my taste. Even the style of scholars has changed in a way not altogether advantageous. In the 19th century scholars from the *Geisteswissenschaften* jumped at each other with a vigour which would shake even a really nasty contemporary, and they did this out of exuberance, not out of a desire to hurt. Dictionaries of remote languages, such as Mediaeval-Latin/English dictionaries gave racy equivalents in English – and so on. Then, slowly, a more measured tone started to insinuate itself, and became the rule. I do not like the change, and I try to restore older ways of writing. In this attempt my guides have

---

wounding comments about people who are in no position to defend themselves. Indeed, Professor Feyerabend seems to insist on the idea that success or power must go to those who have the least respect for consistency and truth in the pursuit of some kind of exploitative paradise of pleasure.'

A touching sermon – but does it concern my book?

Two kinds of comment are made, one about style, one about content. Let us look at the latter to get some preparation for judging the former.

Harré assumes I support political anarchism; I explicitly reject it (cf. the text above and n. 9 of Chapter 2). Harré says I 'dismiss' Lakatos 'out of hand' – but I devote an entire chapter to him. Harré says I 'dismiss' Lakatos for his 'dependence on rational criteria of choice' while I criticize him for the failure of his philosophy to provide such criteria. Harré says I claim 'total licence for myself' (and Galileo) while I suggest that the actions of everyone, scientists, bishops, politicians, comedians be subjected to the judgement of democratic councils. Harré says I recommend inconsistency while I say that the rationalist cannot avoid it. 'My intention is not to replace one set of general rules by another such set' I say on page 32 of *AM* – to no avail. Harré insinuates that I am aiming at an 'exploitative paradise of pleasure' while I want to remove the ideological and financial exploitation of common citizens by a small gang of power- and money-hungry intellectuals (cf. Chapter 2, n. 13 as well as Section 4 of Chapter 3). This last charge, incidentally, which I find again and again is most interesting. It shows a curious attitude towards pleasure: the fact that I am for pleasure seems to count against me. In the issue between Truth and Pleasure Truth is quite obviously regarded as the more important thing. Why? Nobody has an answer. It also shows that intellectuals feel 'exploited' whenever there is the slightest threat that their privileges may be removed and their equality with other citizens restored. What, after all, do I suggest? I suggest that intellectuals be *used*, *praised*, *paid* but *not* permitted to shape society in their image. If this is exploitation, then one has to make the best of it. However that may be – Harré's reading abilities certainly are not very highly developed.

Which brings me to the matter of style. As in the case of Agassi I have spent quite some time looking for the comments 'about women, friends and colleagues, indeed about everyone likely to place trust in someone else' that have so upset Harré. I could not find them. Am I blind or is he hallucinating? It must be the latter considering his inability, just commented upon to understand what I have written and considering also that not too long ago he compared Popper's style with that of GBS. Small wonder that to an eye like that I must appear as a proponent of 'total licence'.

been journalists and poets such as Brecht (his superbly written theatrical criticism of his youth, not the more weighty notes of his later years), Shaw, Alfred Kerr or, to move back to even earlier times, humanists such as Erasmus and Ulrich von Hutten (not to mention Luther who once called Erasmus *flatus diaboli* – and this was quite in line with the good manners of the time). I have no argument for this preference of mine, I just state it as an idiosyncrasy. I state it because the emotion behind a sentence ('loathing' for example, or the absence of it) can be judged correctly only if one first knows the style in which the sentence is written.

Now we finally come to some differences *in substance* that exist between you and me. What are these differences?

To answer the question, I shall quote a footnote from an earlier version of *AM* that was published in Vol. iv of the *Minnesota Studies for the Philosophy of Science* (Minneapolis 1970). I omitted the footnote (and other material such as a chapter on Mill and Hegel) from the book in order to make room for Imre Lakatos's reply (which now, unfortunately, will never be published). I write:

'The possibilities of Mill's liberalism can be seen from the fact that it provides room for any human desire, and for any human vice. There are no general principles apart from the principle of minimal interference with the lives of individuals, or groups of individuals who have decided to pursue a common aim. For example, *there is no attempt to make the sanctity of human life a principle that would be binding for all.* Those among us who can realize themselves only by killing their fellow human beings and who feel fully alive only when in mortal danger are permitted to form a subsociety of their own where human targets are selected for the hunt, and are hunted down mercilessly, either by a single individual, or by specially trained groups (for a vivid account of such forms of life see the film *The Tenth Victim* which, however, turns the whole affair into a battle between the sexes). So, whoever wants to lead a dangerous life, whoever wants to taste human blood will be permitted to do so within the domain of his own subsociety. *But he will not be permitted to implicate others who are not willing to go his way*; for example, he will not be permitted to force others to participate in a "war of national honour", or what have you. He will not be permitted to cover up whatever guilt he may feel by making a potential murderer out of everyone. It is strange to see how the *general* idea of the sanctity of human life that would object to the formation of subsocieties such as the one just described and that frowns upon simple, innocent and rational murders does not object to the murder of people one has not seen and with whom one has no quarrel. Let us admit that we have different tastes; let those who want to wallow in blood receive the opportunity to do so without giving them the power to make "heroes" of the rest of society. As far as I am concerned, a world in which a louse can live happily is a better world, a more instructive world, a more mature world than a world in which a louse must be wiped out. (For this point of view see the work of Carl Sternheim; for a brief account of Sternheim's philosophy, see Wilhelm Emrich's preface to C. Sternheim, *Aus dem Buergerlichen Heldenleben* Neuwied: Hermann Luchterhand 1969, pp. 5–19). Mill's essay is the first step in the direction of building such a world.

It also seems to me that the United States is very close to a cultural laboratory in the sense of Mill where different forms of life are developed and different modes of human existence tested. There are still many cruel and irrelevant restrictions, and excesses of so-called lawfulness threaten the possibilities which this country contains. However, these restrictions, these excesses, these brutalities occur in the *brains* of human beings; they are not found in the *constitution*. They can be removed by propaganda, enlightenment, special bills, personal effort (Ralph Nader!) and numerous other legal means. Of course, if such enlightenment is regarded as superfluous, if one regards it as irrelevant, if one assumes from the very beginning that the existing possibilities for change are either insufficient, or condemned to failure, if one is determined to use "revolutionary" methods (methods, incidentally, which real revolutionaries such as Lenin have regarded as utterly infantile – see his *Left Wing Communism, an Infantile Disorder* – and which must increase the resistance of the opposition rather than removing it), then the "system" will appear much harder than it really is. It will appear harder, *because one has hardened it oneself*, and the blame falls back on the bigmouth who calls himself a critic of society. It is depressing to see how a system that has much inherent elasticity is increasingly made less responsive by fascists on the Right and extremists on the Left until democracy disappears without having ever had a chance. My criticism, and my plea for anarchism is therefore directed *both* against the traditional puritanism in science and society *and* against the "new", but actually age old, antediluvian, primitive Puritanism of the "new" left which is always based on anger, on frustration, on the urge for revenge, but never on imagination. Restrictions, demands, moral arias, generalized violence everywhere. A plague on both your houses!' So far part of n. 49 of my essay of 1970 (remember that the Vietnam War and the Student 'Protests' were then still very much in existence).

I think you will admit that the society described in this passage has little in common with 'totalitarian China'. Even during the period of the Hundred Flowers the freedom achieved in China was but a fraction of what I think is possible and desirable. Note also, that there is no complete licence. Not all actions are permitted, and a strong police force prevents the various subsocieties from interfering with each other. But as regards the nature of these societies, 'anything goes', especially in the field of education. And with this I come to a further disagreement between you and me. *I* say that the educational institutions of a democracy should in principle teach any subject, *you* say that only 'a knave and a fool' would suggest to introduce Voodoo and astrology to 'State Colleges and Universities'. So, let us take a closer look at the matter.

As far as I am concerned, the situation is childishly simple.

'State Colleges and Universities' are financed by taxpayers. They are therefore subjected to the judgement of the taxpayers *and not* to the judgement of the many intellectual parasites who live off public money.[5] If the taxpayers of California want their universities to teach Voodoo, folk medicine, astrology, rain dance ceremonies, then this is what the universities will have to teach (the *State* universities; private universities such as Stanford university may still continue teaching Popper and von Neumann).

Would taxpayers perhaps be better advised to accept the judgement of experts? They would not, and for obvious reasons.

First, experts have a vested interest in their own playpens, and so they will quite naturally argue that 'education' is impossible without them (can you imagine an Oxford philosopher, or an elementary particle physicist arguing himself out of good money?)

Secondly, scientific experts hardly ever examine the alternatives that might come up in the discussion with the care they take for granted when

---

[5] I welcome most enthusiastically the Baumann amendment which recommends congressional veto power over the 14,000 odd grants the National Science Foundation awards every year. Scientists were very upset by the fact that the amendment was passed by the House of Representatives and the director of the National Academy spoke darkly of totalitarian tendencies. The well paid gentleman does not seem to realize that totalitarianism means direction of the many by the few while the Baumann amendment goes exactly in the opposite direction: it suggests examining what the few are doing with the millions of public money that are put at their disposal in the vain hope that the public will eventually profit from such generosity. Considering the narcissistic chauvinism of science such an examination would seem to be more than reasonable. Of course, it should be extended beyond the narrow limits of supervision of the NSF: every department at a state university must be carefully supervised lest its members use public money for working out their private fantasies under the heading of philosophical, psychological, sociological 'research'.

a problem in their own field is at stake. They agonize over different scientific approaches to the problems of space and time but the idea that the Hopi Genesis might have to add something to cosmology is at once rejected out of hand. Here scientists and, for that matter, all rationalists act very much like the Roman Church acted before them: they denounce unusual and extraordinary views as Pagan superstitions, they deny them every right to make a contribution to the One True Religion.[6] Given power, they will suppress Pagan ideas *as a matter of course* and replace them by their own 'enlightened' philosophy.

Thirdly, the use of experts would be allright if they were only taken from the proper field. Scientists would laugh their heads off (or, to be more realistic: they would be very indignant) if one asked a faith healer and not a surgeon about the details of an operation: obviously the faith healer is the wrong person to ask. But they take it for granted that an astronomer and not an astrologist should be asked about the merits of astrology, or that a Western physician and not a student of the *Nei Ching*   ↬ should decide about the fate of acupuncture. Now – and with this I come to the fourth point – such a procedure would be unobjectionable if the astronomer, or the Western physician, could be assumed to know more about astrology, or acupuncture than the astrologist, or the traditional Chinese doctor. *Unfortunately, this is only rarely the case.* Ignorant and conceited people are permitted to condemn views of which they have only the foggiest notion and with arguments they would not tolerate for a second in their own field. Acupuncture, for example, was condemned not because anyone had examined it, but simply because some vague idea of it did not fit into the general ideology of medical science or, to call things by their proper name, because it was a 'Pagan' subject (the hope for financial rewards has in the meantime led to a considerable change of attitude, however).

What is the effect of this procedure?

The effect is that scientists and 'liberal' rationalists have created one of the most unfortunate embarrassments of democracy. Democracies *as conceived by liberals* are always embarrassed by their joint commitment to 'rationality' – and this today means mostly: science – and the freedom of thought and association. Their way out of the embarrassment is an abrogation of democratic principles where they matter most: in the domain of education. Freedom of thought, it is said, is OK for grownups

---

[6] Such an attitude is frequently found in Galileo. He argues with his fellow mathematicians, he has only contempt for the mathematically uneducated 'rabble' (his own words).

who have already been trained to 'think rationally'. It cannot be granted to every and any member of society and especially the educational institutions must be run in accordance with rational principles. In school one must learn what is the case and that means: Western oriented history, Western oriented cosmology, i.e. science. Thus democracy *as conceived by its present intellectual champions* will never permit the complete survival of special cultures. A liberal–rational democracy cannot contain a Hopi culture in the full sense of the word. It cannot contain a black culture in the full sense of the word. It cannot contain a Jewish culture in the full sense of the word. It can contain these cultures only as secondary grafts on a basic structure that is constituted by an unholy alliance between science, rationalism, and capitalism. This is how a small gang of so-called 'humanitarians' has succeeded in shaping society in their image and in weeding out almost all earlier forms of life.[7] Now this would have been a laudable undertaking if the beliefs constituting these forms of life had been examined with care, and with due respect for those holding the beliefs, and if it had been found that they are a hindrance to the free development of humanity. *No such examination has ever been carried out* and the few individuals who have started taking a closer look at the matter have come to a very different conclusion. What remains in the end behind all the humanitarian verbiage is the white man's assumption of his own intellectual superiority. It is this high handed procedure, this inhumane suppression of views one does not like, this use of 'education' as a cub for beating people into submission which has prompted my contempt for science, rationalism, and all the pretty phrases that go with it ('search for truth'; 'intellectual honesty' etc. etc.: intellectual honesty, my foot!) and not a mythical astronomical disappointment at an early age as you, dear Joske, seem to believe. And I do not see why I should be polite to tyrants, who slobber of humanitarianism and think only of their own petty interests.

There are many more things I would like to say, but a review is short, and the review of a review is even shorter. So, let me conclude with a personal story. For the past half year I have been losing weight, about 25 pounds by now, I got double vision, stomach cramps, I fainted in the

---

[7] The history of the American Indians is a case in point. The first wave of invaders came to enslave them and to 'teach them Christian manners' as it reads in Alexander VI's bull on the new islands and the new continent. The second wave of invaders came to enslave them and to teach them Christian manners of a different kind. By now they have been robbed of all material possessions and their culture has almost disappeared – 'and rightly so' say the rationalists, 'for it was irrational superstition'.

streets of London and felt generally miserable. Naturally, I went to a doctor. The general practitioners (this was in England) did not do me much good. I went to specialists. For three weeks I was subjected to a battery of tests; I was given X-rays, emetics, enemas and each examination made me feel worse than ever. Result: negative (this is a nice paradox: you are sick; you go to a doctor; he makes you feel worse; but he says that you are well). As far as science is concerned, I am as fit as a fiddle. Not being restricted by an undying loyalty to science I started looking for other kinds of healers and I found there are lots of them. Herbalists. Faith healers. Acupuncturists. Masseurs. Hypnotists. All quacks, according to the established medical opinion. The first thing that caught my attention was their method of diagnosis. No painful interference with the organism. Many of these people had developed efficient methods of diagnosing from pulse, colour of eye, of tongue, from gait, and so on. (Later on, when reading the *Nei Ching* which develops the philosophy behind acupuncture, I found that in China this was intentional: the human body must be treated with respect which means one has to find methods of diagnosis that do not violate its dignity.) I was lucky. The second man I consulted told me I had been severely ill for a long time (and that is true: for the past 20 years I wavered between long periods of health and other periods when I was hardly able to totter along, but without any scientifically detectable signs of illness), that he was going to treat me twice in order to see whether I responded and that he might take me on if I did. After the first treatment I *felt* better than I had felt for a long time and there were *physical* improvements as well, a long-lasting dysentery stopped and my urine cleared up. None of my 'scientific' doctors had been able to achieve that. What did he do? A simple massage which, as I found later, stimulated the acupuncture points of liver and stomach. Here in Berkeley I have a faith healer and an acupuncturist, and I am now slowly recovering.

So, what I discovered was this: there exists a vast amount of valuable medical knowledge that is frowned upon and treated with contempt by the medical profession. We also know, from more recent anthropological work, that 'primitive' tribes possess analogous knowledge not only in the field of medicine, but in botanics, zoology, general biology, archaeologists have discovered the remnants of a highly sophisticated Stone Age astronomy with observatories, experts, application in exploratory voyages which was accepted across cultural lines, throughout the European continent. Myths, properly interpreted, have turned out to be repositories of knowledge unsuspected by science (but confirmable by scientific research, once the matter is taken up) and occasionally in conflict with it. *There is much we can learn and have to learn from our ancient ancestors, and our 'primitive' fellow men.* In view of this situation, must we not say, that our educational policies, including your own, dear Joske, are very ill conceived and narrow minded, to say the least? They are *totalitarian*, for they make the ideology of a small gang of intellectuals the measure of everything. And they are *shortsighted*, because this ideology is severely limited, it is a hindrance to harmony and progress. Let us become more modest; let us admit that Western rationalism is but one of many myths, not necessarily the best, let us adapt our education, and our society as a whole to this modesty and, maybe, we shall be able to return to a paradise which was once our own but which now seems to have been lost in noise, smog, greed, and rationalistic conceit.

All the best
Paul

Postscript 1977

Professor Agassi wrote a reply to my comments which shows no improvement of his reading ability. He reads my *criticism* of liberal democracy as a *recommendation* to Jews to return to the religion of their forefathers, to American Indians to resume their old ways, rain dances included, and he bemoans the 'reactionary' character of the recommendation. Reactionary? That takes it for granted that the step into science, technology and liberal democracy was not a mistake – which is the question at issue. It is also taken for granted that older practices, for example rain dances don't work – but who has examined *that* matter (and note that for examining it we would have to restore the harmony between man and nature that existed before the Indian tribes were broken up and annihilated). Besides, I don't say that Jews, or American Indians *should resume* their old ways, I say that those who *want to resume* them should be able to do so, first, because in a democracy everyone should be able to live as he sees fit and, second, because no ideology and way of life is so perfect that it cannot be improved by consulting (ancient) alternatives.

Agassi asks 'who will bring them' (the Jews, the Indians) 'back to modernity when the therapy session is over?' This repeats the mistake I have just commented upon. I don't want to change people's minds by some imagined therapy, I object to the real therapy, called 'education' that is constantly applied to their children. And if people have decided to resume their old ways – then why should anyone have to bring them 'back to modernity'? Is 'modernity' that marvellous that one has to return to it no matter what insights one has gathered on excursions into different fields?

Also, I cannot accept Agassi's simpleminded way of dealing with evil. 'I mention Dachau and Buchenwald in order to refute the "anything goes" thesis' he writes; 'anything short of these, of course'. Of course? Is that all he has to say? Are we supposed to stop thinking at this point? Are we supposed to accept revulsion (and the cowardly opportunism of those once on the wrong side) as a basis of argument? Or must not a rationalist (which, fortunately, I am not) examine the *credentials* of the revulsion and if he must, then how are these credentials to be found? When Remigius, the Inquisitor, was an old man he remembered with sadness how in his youth he had saved children of witches from the flames instead of burning them, as was the rule, and how he had in this way contributed to their eternal damnation. Remigius was an honest man and a humanitarian and yet his beliefs about the world and the fate of man made him act in a way which somebody not aware of his motives will 'of course' regard as the very opposite of humanitarianism. 'Of course' many Nazis were puny and despicable men – every new publication including the most recent publication of Goebbels' diaries shows that – and of a very different calibre from Remigius. But puny and despicable men are human, they have been created in the image of God and that alone requires of us to treat them with greater circumspection than on the basis of a mere 'of course'. For a long time now I have considered writing a play about such a despicable character. He is introduced – and we at once loathe him with all our heart. He acts, and our loathing increases. But as the play proceeds we come to know him better. We see how his actions flow from his humanity, from his full and genuine humanity and not from some decayed part of it. He is no longer an outcast of mankind, he is part of it, though a puzzling part. Moreover, we not only understand his actions as human actions, we perceive an inherent reason and begin to get attracted by it. We realize that we might act exactly as he does and we already *want* to act in that manner. We are on the way to becoming him and acting like him. What kind of person do I have in mind? He may be an SS officer, he may be an Aztec engaged in ritual killing or self mutilation, he may be a rationalist habitually killing *minds*[8] – take your pick. And, finally, the play resumes its initial position and our loathing returns.

[8] Having seen hundreds of unsmiling and narrowminded young people moving grimly along the Path of Reason (critical reason, dogmatic reason – that does not make any difference) I ask myself what kind of culture it is that has eulogies, prizes, respect for such killing of *souls* while turning with standardized revulsion from the killing of bodies. Is not the soul more important than the body? Should not the same or an even greater punishment be extended to our 'teachers' and our 'intellectual leaders' than is now extended to individual and collective murderers? Should not guilty teachers be found out with the same vigour one

Such a play, I think, is entirely possible (the film could be used even more effectively). It would fail with people who are frozen by ideology but to most people it would demonstrate that to be human is to be evil as well as good, rational as well as irrational, divine as well as loathsome, that it is possible to be good while being evil and evil while attempting to be good. The human race is like the whole realm of nature – there is not one item of behaviour it is not capable of duplicating.

In these circumstances – what will be our attitude towards Dachau and Buchenwald? I do not know. But one thing is certain: Agassi's elegant 'of course' shelves problems we must consider and live through if we want fully to realize our humanity. I am not aware that anyone has done this so far. Which accounts for the vapid nature of almost everything that has been written on the matter.[9]

---

applies to the hunting of Nazi octogenarians? Are not the so-called 'leaders of mankind' – men such as Christ, Buddha, St Augustine, Luther, Marx, some of our greatest criminals (it is different with Erasmus, or Lessing, or Heine). All these questions are pushed aside with the facile 'of course' that freezes standard reactions instead of making us *think*.

[9] An exception is Hannah Arendt's marvellous book on Eichmann and the banality of evil. Cf. also the beginning of Section 11 of Part Two of the present book.

Chapter 2

Logic, Literacy and Professor Gellner

It is always a pleasant surprise for an author to find a critic who understands his philosophy, agrees with it, and shows some ability to develop it further. It is even more gratifying to meet a thinker who shares with him not just ideas but some other idiosyncracy, especially when the idiosyncracy is unpopular, and frowned upon by the profession. For years Lakatos and I were alone in our attempt to inject a little life, some personal note into philosophical debate. Since Imre's death nobody was left to support me in this enterprise. Now the review of my book in this *Journal*[1] introduces a writer who is not only willing to leave the narrow path of academic prose and reasoning but who has great talent in this direction, who is a master in the art of invective and who has considerably augmented the inventory of rhetorical techniques. Perhaps I should have been grateful for the support which my efforts seemed to receive from such unexpected quarters and should not have probed further, but unfortunately my pedantry got the better of my gratitude. I soon discovered that though the reviewer writes well he does not at all write correctly. His ability to give colour to his own ideas and impressions is marred by a striking blindness for the ideas, the motives, the procedures of others. His interpretations of my text are but rarely conscious distortions, as they would be with a sophisticated rhetorician, they are mostly simple errors of reading and comprehension. Indeed, I discovered that we have here not a well planned extension of the art of *rhetorical* argument, but side effects of an abortive attempt at a *rational* criticism. This being the case I unfortunately cannot praise Gellner for his rhetorical acumen but am reduced to the unpleasant task of enumerating trivial errors and misunderstandings. In the remarks that follow I shall try to sweeten this task

[1] Ernest Gellner, 'Beyond Truth and Falsity', *British Journal for the Philosophy of Science*, Vol. 26, 1975, pp. 331–42. My reply was first published in the *British Journal for the Philosophy of Science*, Vol. 27, 1976; I have made some changes and added a few lines here and there.

as well as I can, both for myself, and for my readers. I shall concentrate on points that not only reveal Gellner's procedure but are also of general interest and whose discussion, one hopes, will achieve a little more than a mere return to the text from which Gellner started.

Gellner's review contains (*i*) a presentation of my main theses and arguments; (*ii*) a criticism of my style and an evaluation of its results; (*iii*) a sociological analysis of the 'happening' that is my book. I shall now discuss these points in order.

(*i*) At first sight it would seem that Gellner gives a fairly accurate presentation of what I say for the sentences he writes down resemble sentences that occur in my book. Now in my book the sentences are either part of a context that contains qualifications, or they describe views I do not hold. Read with these qualifications or intentions in mind, they correctly express my argument. Gellner does not consider the qualifications and he proceeds as if I were stating my own opinions throughout. The *prima facie* correctness therefore hides some considerable errors.

Take sentence (1) in his 'spinal column' – 'the actual history of science shows that the real advances of knowledge contradict all available methodologies' (p. 333) – this is supposed to be a thesis that is held, or insinuated in my book. According to Gellner it is 'clearly the nucleus from which the whole thing grew'. To the unwary reader the statement suggests (*a*) that I claim to know the truth of some historical facts and generalizations, (*b*) that I presume to possess insight into the even more difficult problem as to what counts as an advance of knowledge and (c) that I refute norms by facts. This is not an abstract possibility. Gellner himself ascribes to me claim (*a*), uses the ascription to accuse me of incoherence (p. 337) and explains my confidence in the face of such incoherence by calling my argument a 'game which [I] cannot lose' (p. 334). But (1) interpreted as entailing (*a*), (*b*), (*c*) is not a thesis I hold. I do not say that methodologies fail because they are merely contradicted by facts – arguments of this kind have been shown to be of questionable value long ago – I say they fail because applied in the circumstances enumerated in my case studies they would have hindered progress. Nor do I claim to possess special knowledge as to what constitutes progress,[2] I simply take the cue from my opponents. *They* prefer Galileo to Aristotle. *They* say that the transition Aristotle → Galileo is a step in the right direction. I only add that this step not only *was not achieved*, but *could not have been achieved* with the methods favoured by them. But does not

[2] *AM*, p. 27.

this argument involve highly complex statements concerning facts, tendencies, physical and historical possibilities? Of course it does, but note that I am not committed to asserting their truth as Gellner surmises. My aim is not to establish the truth of propositions, my aim is to make my opponent change his mind. To achieve this I provide him with statements such as 'no single theory ever agrees with all the known facts in its domain.[3] I use such statements because I assume that being a rationalist he will be affected by them in a predictable way. He will compare them with what he regards as relevant evidence, for example, he will look up records of experiments. This activity, combined with his rationalistic ideology will in the end cause him 'to accept them as true' (this is how *he* will describe the matter) and so he will perceive a difficulty for some of his favourite methodologies. But don't I now even make more far reaching assumptions about the minds of people, the structure of records, the changes which occur in the former when they are confronted with the latter? Quite right, but these assumptions are not part of my argument with the reader. They are part of an argument I carry on with myself and which concerns the efficiency of my persuasion. The structure of this latter discourse is of no concern for the rationalist who, after all, insists on separating the 'objective content' of an argument from its 'motivation'. All he *needs* to consider, all he is *permitted* to consider is how the statements surrounding the case studies in my book are related to each other and to the historical material and whether they can be read as an argument in *his* sense. I admit that my procedure succeeds by *manipulating* the rationalist but note that I manipulate him in a way in which he *wants* to be manipulated and constantly manipulates *others*: I provide him with material which interpreted in accordance with the rationalistic code creates difficulties for views he holds. Do *I* have to interpret the material as he does? Do *I* have to 'take it seriously'? Certainly not, for the motivation behind an argument does not affect its rationality and is therefore not subjected to any restriction.

Gellner's (2) and (3) and the reasons he says I give for them are equally inadequate. I would not accept the 'this shows' (p. 333) for I know that we may 'improve these methodologies' (p. 334); I would not accept the 'all', especially as I have formulated what I believe to be sensible methodological suggestions[4] and argue only against *universal* methods which abstract both from the content of a theory and from the context of the

---

[3] *ibid*, p. 55.
[4] For the use of *ad hoc* hypotheses see *AM*, pp. 178 and 97; of a plurality of theories p. 41; of counter induction, Chapter 6; 'backward movements', p. 153; connections with influential

debate.[5] Nor would I ever presume to *legislate* for scientists or, for that matter, for anyone else as Gellner implies in (5) and (6). I did this in earlier papers when I was younger, more ignorant, more pushy and considerably more conceited.[6] At that time my arguments for proliferation were indeed designed to show that a monistic life is not worth living and they urged everyone to think, feel, live through a competition of alternatives. Today the same arguments are offered with a very different *purpose* in mind, and they lead to a very different *result*.[7] Scientists and rationalists have by now almost succeeded in making their views the basis of Western Democracy. They concede, though with extremely bad grace, that other ideas may be *heard*, but they would not permit them a role in the planning and the completion of fundamental institutions such as law, education, economics. Democratic principles as they are practiced today are therefore incompatible with the undisturbed existence, development, growth of special cultures. A rational-liberal democracy cannot contain a Hopi culture in the full sense of the word. It cannot contain a Black culture in the full sense of the word. It cannot contain a Jewish culture in the full sense of the word. It can contain these cultures only as *secondary grafts* on a basic structure that is constituted by an unholy alliance of science, rationalism (and capitalism). All attempts to revive traditions that were pushed aside and eliminated in the course of the expansion of Western

---

ideologies p. 193; or other refuted theories p. 142; use of political force to revive theories that are 'scientifically untenable' p. 50; skipping over difficulties, Appendix 2 – and so on. Cf. also my comparison of logical and anthropological procedures for the discovery of methodological rules p. 260.

[5] *AM*, p. 295, 'fixed and universal rules'. Gellner's reference to the rules for solving quadratic equations (p. 334) is therefore beside the point (and the example is much more complex than he seems to assume). Furthermore, introducing what he calls his 'own view of the matter' just shows how little he has understood mine: there is no difference between the two.

[6] Gellner does Popper an injustice when connecting this immature earlier work with the 'Popperite movement' (p. 332). True, there are some acknowledgments to Popper in these papers, but they are friendly gestures, not historical statements (I also mention my girl friends). True, some points sound quite Popperian to somebody who has only read Popper, but they stem from Mill, Mach, Boltzmann, Duhem and above all, Wittgenstein. True, in my book I occasionally make fun of Popper (I 'lash out' says Gellner – p. 332 – who does not seem to be able to distinguish mockery from aggression), but only to tease Lakatos (p. 8) who was supposed to reply to my book and who was overly impressed by Popper, and not because of 'excessive demands for conformity and involvement made by the master' (*ibid.*, 'master' indeed!).

[7] The change is due to a conversation that occurred at Professor von Weizsaecker's seminar in Hamburg 1965: Professor von Weizsaecker gave a detailed account of the Copenhagen interpretation and showed how it could be applied to specific problems. I pressed the point that hidden variable theories are needed to increase the empirical content of the orthodox view when I suddenly realized how barren such an attitude is in the face of concrete research.

culture and to make them the basis of existence for special groups run into an impenetrable stone wall of rationalistic phrases and prejudices. I try to show that there are no arguments to support this wall and that some principles implicit in science definitely favour its removal.[8] There is no attempt on my part to show 'that an extreme form of relativism is *valid*' (p. 336), I do not try to *justify* 'the autonomy of every mood, every caprice, and every individual' (*ibid*.), I merely argue that *the path to relativism has not yet been closed by reason* so that the rationalist cannot object to anyone entering it. Of course, I have considerable *sympathy* for this path and I think it is the path of growth and freedom, but that is a different story.

More specifically, the situation is as follows. I do not show that proliferation *should be used*, I only show that the rationalist *cannot exclude it*. And this point is not made negatively, by showing how existing objections break down but *via* an argument that derives proliferation from the monist's own ideology. The argument is in two parts, one based on science, the other on the relation between scientific and non scientific ideologies. The argument from science says that proliferation follows from the scientist's own demand for high empirical content (*AM*, pp. 41 and 47). I do not accept the demand which is only one of the many ways to bring order into our beliefs (*ibid*. p. 204) and so I do not argue for its implications. What I do say is that scientists who are in favour of high empirical content are committed to proliferation as well and therefore cannot reject it.[9] The argument from the existence of incommensurable ideologies says

---

[8] Gellner says that the social consequences of rationalism (or irrationalism) are 'tangential' to my main concerns (p. 339). *The very opposite is the case*. For me democracy, the right of people to arrange their lives as they see fit comes first, 'rationality', 'truth' and all those other inventions of our intellectuals come second. This, incidentally, is the main reason why I prefer Mill to Popper and why I have nothing but contempt for the sham-modesty of our critical rationalists who fall over themselves in their concern for 'freedom' or an 'Open Society' but start erecting obstacles whenever people want to live in accordance with the traditions of their forefathers.

[9] I explicitly state that 'my intention is not to replace one set of general rules by another such set: my intention is, rather, to convince the reader that *all methodologies, even the most obvious ones, have their limits*' (p. 32 my italics) – which does not stop Gellner from presenting 'pointless proliferation of viewpoints' (p. 340) as a *positive* doctrine of mine and explaining recalcitrant passages either by some hidden incoherence on my part (p. 333) or by my 'clowning' (p. 338).

At first sight this looks as if Gellner had made an interesting contribution to the art of rhetorical argument: assume you are supposed to review a book the major part of which is beyond your comprehension (p. 334). You concentrate on the 'rest of the book' (*ibid*.), say that 'it is of some interest' (*ibid*.) and extract a series of theses from it. You also present the theses in a systematic way, you arrange them in a 'spinal column' (p. 333) and add arguments to show that you are proceeding in a fair and rational manner. If the arguments or the theses contradict statements found in the book, you accuse the author of incoherence. If the author

(*a*) that their comparison does not involve content and so cannot be in terms of truth *vs.* falsehood, except rhetorically (Chapter 17),[10] (*b*) that every ideology possesses methods of its own and that the comparative evaluation of methods has not even started. All we have is the dogmatic belief in the excellence of the 'methods of science' (where everyone has different ideas as to what these methods are). But, (*c*), non-scientific views and methods, far from being complete failures, have led to amazing discoveries in the past, they are often better than the corresponding scientific views and have better results (cf. *AM*, pp. 49ff.).[11] Taking all these arguments together I conclude that a person who wants to introduce unusual views, methods, forms of life or who wants to revive such reviews, methods, forms of life *does not need to hesitate* for reason has not yet succeeded in putting any obstacles in his path and scientific reason even urges him to increase the number of alternatives. The only obstacles he is going to meet are prejudice and conceit.

Let us stay a little longer with this matter of proliferation to get a good look at Gellner the reviewer in action. We have seen that Gellner misjudges the *role* of proliferation in my arguments. He does not understand its *consequences* either. He chides me for the 'welcome admission' (p. 339) that technology cannot be had without scientists. To start with, there is no such admission on my part. I address people who fear that the separa-

---

is not your run-of-the-mill academic square but makes a joke now and then you can explain the conflict by his 'clowning' as well. So you can have your cake – you don't have to understand everything you read – and eat it, too – you can write a vigorous, decisive, witty review. Unfortunately Gellner is only dimly aware of what he is doing. Most of the time he thinks he is producing a straightforward rational criticism (*passim*). So we cannot praise him for his rhetorical acumen but are reduced to noting his inability to understand what he reads.

[10] It is therefore not correct to describe my view as implying that 'almost everything might contain some truth' (p. 335). Gellner's 'Epistemological theories . . . give us some insight in how to choose between whole styles of thought' (p. 336) does not solve the conundrum either for every self respecting 'style of thought' will of course have its own epistemology (cf. *AM*, p. 246). The principle, however, that a 'culture which subjects its cognitive capital to testing by arbiters *who are not under its own control*' is superior to one 'which does not do so' (p.336) would prefer cultures with oracles to cultures with scientific experiments for the latter are on the whole much more strictly controlled than the former.

[11] Gellner is 'sceptic[al] about [such] truly amazing achievement[s]' (p. 242) and this is understandable, for he does not know the literature. What he does know is that most of his readers share his scepticism and will be impressed by a statement of it. Had he also known that they are sceptical because they are as illiterate as he then we could have congratulated him on his elegant use of a fine rhetorical principle: if your opponent assumes matters your readers are not likely to have heard about then score a point by acting as if these matters did not exist and as if your opponent was talking through his hat. But Gellner thinks his information is complete which means: we are not dealing with rhetorical sophistication but with ignorance, pure and simple.

tion of state and science will lead to a breakdown of health, public transportation, radio, T.V. and so on because – and this is *their* reason, not mine – technology cannot exist without scientists (*AM*, p. 299). Trying to alleviate this fear I could either deny the reason, i.e. I could argue that technology does not need closed societies of highly qualified experts to succeed – this I do on page 307, though rather briefly – or I could give an answer that leaves the reason untouched – and this I do on page 299. Assuming that my reader can follow an argument without being constantly reminded of its presuppositions I state the view of my opponent and my own views side by side, as in a dialogue, but without making the sides explicit. Page 299, e.g., means: *Opponent*: but will not a separation of state and science lead to a breakdown of technology? *I*: you seem to believe that technology without experts is impossible, I am rather doubtful about this assertion, but let us take it for granted. Then you must realize that there will always be people who prefer being scientists . . . and so on. Gellner conflates the opponent's thesis and my own reply, turns the conglomerate into a single view, ascribes the view to me, analyses it, and triumphantly exposes its incoherence. And as he starts running different contexts into each other whenever the argument becomes a little complicated, he has now another, and most efficient method of finding inconsistencies in my book. But the 'plot' (p. 338) he unveils in this manner is but a reflection of his own simpleminded reading habits: he understands 'the cat is on the mat'; he may still understand, though with some effort 'Joe says that the cat is on the mat' but 'the cat is on the mat – do you really believe this? I don't' shows to him that the author says that the cat both is and is not on the mat and thus defends incoherence. This is Gellner's third 'contribution' to the art of rhetorical argument.[12]

Secondly, the 'welcome admission' is not contrary to the idea of proliferation. Proliferation does not mean that people can't have well defined and even dogmatic views, it means that research consists in playing

---

[12] Typical example: on p. 21, n. 12 I defend the pacifism of the Dadaists and say that I am against violence. On p. 187 I say that *political*, or *eschatological* anarchism regards violence as necessary. Gellner (p. 340) runs the two passages into one and says that I 'incoherently conflate' (beautiful words!) the 'mystique of violence' with a 'pacifist no-fly-hurting posture' which in turn is combined with 'cognitive/productive parasitism'. Well roared pussycat – but don't you think you should have read the text a little more carefully or let somebody else explain it to you in case you can't read? The text says that violence is necessary *according to political anarchism* and adds that political anarchism is a doctrine I reject. The very first sentence of the book calls political anarchism 'not the most attractive political philosophy' (p. 17) and on p. 189 I again distinguish my views from political anarchism, just to be on the safe side. All in vain.

views off against each other rather than in pursuing a single view to the bitter end. Proliferation does not entail that scientists are *excluded* or that statements like 'we need scientists' or 'Lysenko failed' (p. 341) are *removed* from the domain of debate; it means that statements denying them or making fun of them are *admitted* and even welcome in the hope that this will be of advantage. With Liberalism the situation is exactly the same. Gellner chides me for explaining the difference between Popper and Mill by reference to Popper's puritanism. 'My own liberalism' he says with pride (p. 332) 'is such that I hold that even Puritans are not debarred from truth'. I did not say they were. I said that Popper's liberalism is different from Mill's and that Puritanism is one explanation (another explanation is that Popper never faced a situation that forced him to revise his whole philosophy and is perhaps incapable of recognizing such a situation). Nor does one become illiberal when denying truth to a Puritan. Liberalism, as Gellner ought to know, is a doctrine about *institutions* and not about *individual beliefs*. It does not regulate individual beliefs, it says that nothing may be excluded from the debate. A liberal is not a mealymouthed wishywashy nobody who understands anything and forgives everything, he is a man or a woman with occasionally quite strong and dogmatic beliefs among them the belief that ideas must not be removed by institutional means. Thus, being a liberal, I do not have to admit that Puritans have a chance of finding truth. All I am required to do is to let them have their say and not to stop them by institutional means. But of course I may write pamphlets against them and ridicule them for their strange opinions.

Finally, there is Gellner's remark about 'pointless proliferation' (p. 340). Obviously proliferation is not his cup of tea. But why not a single word about the arguments in Chapters 3 and 4 of *AM* where it is shown that and how proliferation can increase content? (Why not a single word about Mill's excellent arguments for proliferation in *On Liberty*?) Did he regard the arguments as irrelevant? Did he detect flaws? Or is it perhaps the case that arguments of more than two lines exceed his attention span? His remarks about my 'welcome admission' already commented upon suggest the last interpretation. Again illiteracy is found to be the moving force behind Gellner's observations.

To sum up: though I am personally in favour of a plurality of ideas, methods, forms of life, I have not tried to *support* this belief by argument. My arguments are rather of a negative kind, they show that reason and science *cannot exclude* such a plurality. Neither reason nor science are strong enough to impose restrictions on democracy and to prevent people from introducing their own most cherished traditions into it. (Another result is that rationalists have not yet succeeded in defeating scepticism – all views are *equally good* – or its natural extension – *any* evaluation of theories and forms of life is acceptable.[13])

(ii) Rationalists cannot rationally exclude myth and ancient traditions from the basic fabric of democracy. Yet they push them aside, using sophistry, pressure tactics, dogmatic pronouncements many of which they regard as arguments, and *present* in the form of arguments. Such pseudo-reasoning can either be exposed by scholarly analysis, or it can be held up to ridicule. I chose the latter path, partly because I had given arguments where they were necessary, partly because I could not see myself solemnly taking apart these products of conceit and pomposity. Gellner neither likes my procedure, nor does he understand its function. He thinks I use it as a 'criticism evading ploy' (p. 338) when in fact I apply it in a domain where the opponent does much huffing and puffing but is no longer engaged in a rational debate. Having decided to disregard 'the extensive parts [of my book] which argue [this] viewpoint' (p. 333) Gellner was without guidance when entering the domain; being taken in as much as everyone else he did not discover its limits and so, quite naturally, he is upset at what he thinks is unfair and irrational treatment of people 'who ask questions about knowledge in good faith' (p. 342). But the trouble is that this 'good faith' is a faith in principles that are far beyond the reach of argument and are accepted only because of the rationalist's say-so; the trouble is that it is a faith in principles that belong to the theology of rationalism.

Gellner also objects to the use of ridicule and frivolity. '[E]ncouraged by the spirit of the times' he writes (p. 334) both have been 'allowed to intrude within the covers of the book.' They have been 'allowed to intrude' – this means that ridicule and frivolity *are* there, but *should not* be there. Why not? Presumably because they should not occur in books of a certain kind, for example they should not occur in scholarly books Now, first of all, what little bird has told Professor Gellner that I intended to write a scholarly treatise? In my dedication I make it clear that my book was conceived as a *letter* (p. 7) to Lakatos and that its style would be

---

[13] *AM*, p. 189. Gellner ascribes the extension to 'temperamental exuberance' (p. 335) while it is in fact the result of an application of the sceptic's method (of balancing each judgement by its opposite) to his own basic principle (all views are equally good). Here as elsewhere Gellner is quick to explain positions by psychology when they are in fact the results of argument.

that of a letter. (Also, I am not a scholar, and have no wish to be a scholar.[14]) Secondly, why should scholarly books be dry, impersonal, lacking in frivolity?[15] The great writers of the 18th century, Hume, Dr Johnson, Voltaire, Lessing, Diderot who introduced new ideas, new standards, new ways of expressing thought and feelings wrote a lively and vigorous style, they called a spade a spade, a fool a fool and an impostor an impostor. Scholarly debates were still very lively in the 19th century, the number of insults occasionally rivalling the number of footnotes. Dictionaries of recondite languages (mediaeval Latin/English; Sanskrit/English) used racy equivalents, introductions to important editions teemed with ambiguous insinuations. Then, gradually, a more measured tone set in, people became more solemn, they frowned on levity and personal remarks and behaved as if they were playing parts in a strange and highly formalized drama. Language became as colourless and as indistinct as the business suit which is now worn by everyone, by the scholar, by the businessman, by the professional killer. Being accustomed to a dry and impersonal style the reader is disturbed by every deviation from the dreary norm and sees in it an obvious sign of arrogance and aggression; viewing authority with almost religious awe he gets into a frenzy when he sees someone pluck the beard of his favourite prophet. This, my dear Professor Gellner, is the 'spirit of the times' and not the attempt of a few outsiders to restore older and less formalized ways of writing. I do not know how the change came about though I suspect that the 'great men' of today being dimly aware of their dwarfish stature encourage ways of writing that are equally colourless so that by contrast they may still appear to have some signs of life. I do not see any advantage in this procedure and I do not see why I should accept it as a *fait accompli*. Let us now have a look at the explanation Gellner has for my dissent.

(*iii*) According to Gellner I 'conflate incoherently' a 'mystique of violence' with a 'pacifist no-fly-hurting posture' to which I add 'cognitive/productive parasitism' (p. 340).

We have seen how the first part of this change arises. Gellner 'conflates'

---

[14] Gellner says history and philosophy of science are 'area[s] of [my] professional expertise'. They are not, as every historian and philosopher of science will be most happy to confirm. Besides – how would *he* know?

[15] Like Gellner, Rom Harré (*Mind* 1977) speaks of the 'misplaced trendiness' of my book. 'Misplaced'? That assumes that *AM* is not the place for 'trendiness' to occur. But *AM* is not a scholarly book; it is a pamphlet, a letter to a friend of mine who enjoyed a vigorous debate and was going to reply to it in equal terms. And as regards the charge of 'trendiness' it only means that I don't follow Harré's own trend of heaping praise on minor lights of a dying profession and so writing in an impersonal and anaemic style. Cf. pp. 130–1, n. 4.

passages which express my own views with passages describing the views of others. The incoherence is in his reading, it is not in my text. The second part of the charge is a puzzle to me and I can only explain it by some scientistic tendencies on the part of Gellner, or by some amazing lack of reading skills. On page 300, which Gellner quotes, though incompletely, I say that scientists may have some interesting ideas and gadgets to offer, that we should listen to their ideas and use their gadgets but without permitting them to build society in their own image, e.g. without permitting them to become the masters of education: there should be a separation of state and science just as there is now a separation between state and church. The reason for the separation is simple: every profession has an ideology and a drive for power that goes far beyond its achievements and it is the task of democracy to keep this ideology and this drive under control. Science is here no different from other institutions as can be seen from the attitude of official medicine towards unusual ideas that have not passed through its own channels (and note that the comparative efficiency of these channels has never been examined in the past and that research that is being carried out today has revealed appalling shortcomings). Calling such a procedure 'cognitive parasitism' is as sensible as calling parasites all those astronomers who consult ancient records but without adopting the theology that played an essential part in their construction and interpretation. And as regards the 'productive' side of this 'parasitism' I need only repeat that scientists will of course be amply rewarded for their services[16] – which is more than is granted to the taxpayer who is supposed to finance scientific research but without any assurance that his needs will be taken into account.[17]

---

[16] Gellner omits the passage describing the rewards from his quotation which shows that his literacy varies from place to place. Occasionally he just does not understand a word of what he is reading. But on other occasions he understands very well, and then he changes the text: he is either illiterate or a liar.

[17] The charge of cognitive/productive parasitism stands the actual situation on its head. What is a parasite? A parasite is a man or a woman who gets something for nothing. Today many scientists and intellectuals are parasites in precisely that sense. They get something – large salaries, expensive playpens – for nothing. For let us not forget that only a fraction of the research and the teaching that is being carried out at state universities and other tax-supported institutions such as the National Science Foundation benefits the community at large, or is even conceived with the idea of such benefits. Even research that seems eminently practical is conducted in a way that decreases the chance of speedy practical results: one does not explore successful procedures that are theoretically opaque, one prefers an approach that gives 'understanding' where the criteria of understanding are defined by the researchers themselves, e.g. cancer research. Alternative procedures are rejected without examination not because they are faulty, but because they clash with the equally unexamined beliefs of one's own sect. The attitude has deplorable consequences in education. Valuable traditions

Having stated his version of my views, Gellner starts classifying them. How does he proceed? He has heard that I am now in Berkeley and that some people in Berkeley some time ago preached peace but were attracted by violence. 'Putting two and two together' he calls my 'doctrines about violence' (which, as we have seen, are doctrines I describe and reject) 'Californian' which will hardly please Ronald Reagan and his numerous followers all over the state, from Los Angeles, Orange County all the way up to Goose Lake and does injustice to the many fine and upstanding revolutionaries that arose at the LSE. Next Gellner remembers that I was born in Vienna and he has also the idea, no doubt from some American movies, that the Viennese like to live a relaxed life. Again 'putting two and two together' he calls some of my suggestions 'characteristically Viennese' (p. 332) which is just as sensible as calling Popper's dogmatism 'Popish conceit' on the grounds that Popper is from Vienna and that Vienna is full of Catholics. One sometimes wonders whether Gellner took these explanations seriously or whether they are not rather attempts at rhetoric when reason fails him. A rhetorician, of course, would know that invective is successful only if its main elements are based on fact and if it does not offend the mind of the reader. So it all depends for whom Gellner wrote the review. I hear that Gellner's colleagues at the London School of Economics were very happy with it – he seems to have correctly gauged their level of intelligence. For more critical readers, however, the review is just another example of the fact that intellectuals remain rationalists (or 'critical' rationalists) only as long as it suits their purpose.

This brings me to the last point of my reply, viz. Gellner's attempt to defend Lakatos. Lakatos, Gellner says, 'observed the highest standards of rigour, lucidity and responsibility' both in 'his writings and [in his] lectures' (p. 332). Poor Imre! If one thing is certain it is that the standards

---

are eliminated, the lives of people are impoverished not because the traditions have been shown to be inadequate but because they don't agree with the basic assumptions of science and because scientists have now the power to impose their ideology on almost everyone. Thus scientists and intellectuals are not just parasites of the pocketbook, they are parasites of the mind as well and they will continue in their path unless democracy puts them in their place. This is what I suggest. I suggest that there be a careful examination of the way in which scientists use public money and of the doctrines they impose upon the young. I suggest that promising research be properly rewarded – but the underlying ideology will not automatically become part of basic education (the ideology of prison guards may be excellent for keeping prisoners in their place, but it may be entirely unsuited as a basis for general education). Considering the narcissistic chauvinism of science such an examination would seem to be more than reasonable. It takes a sociologist to describe the procedure as 'cognitive/productive parasitism', but it is understandable why Gellner reacts in this way: realization of my plan means the end of a comfortable life for him.

of Lakatos were very different from those of Gellner. Lakatos was not at all averse to tricks, frivolity, ridicule in his lectures, he never descended to the *tierischer Ernst* which is Gellner's basic attitude, despite all his attempts at levity, and even in his writings he more than once left the path of rational argument to take a well aimed swipe at his opponent. On the other hand, Lakatos could read, and he had no use for the types of explanation I have just discussed and which are found not only in Gellner's review, but in all his works. Nor was it necessary to warn the reader of my dedication (p. 331). Imre Lakatos, whom I asked permission accepted it with glee; he knew that it was a joking reference to Chapter 16 that deals with his views and he looked forward to writing his reply. This reply, I am sure, would not have consisted in the bald statement that 'the claim that Lakatos' position was an 'anarchism in disguise' is totally unfounded (p. 331), for he paid attention to my arguments and thought that I had made a good case for this claim. Read it as you will, Gellner's praise of Lakatos and his attempt to defend him from me is an undeserved insult to the memory of a great scholar and a marvellous human being.

Chapter 3

Marxist Fairytales from Australia

Sydney has one opera house, one Arts centre, one zoo, one harbour, but two philosophy departments. The reason for this abundance is not any overwhelming demand for philosophy among the antipodes but the fact that philosophy has party lines, that different party lines don't always get on with each other and that in Sydney one has decided to keep peace by institutional separation. The two members of the Department for General Philosophy who have reviewed my book *Against Method* clearly take their own party line very seriously indeed. They notice that '*AM* appear[s] under the imprint of one of the leading publishers of English-language books of a "left" (mostly Marxist) orientation', they have heard that my views 'have found some reception among Marxist and radicals generally', they are concerned that good, honest, sober Marxists might be led astray, and so they have decided to review my book 'from a Marxist point of view'.[1] Their result is that I am a sheep in wolf's clothing: I have failed to transcend the ideology I am trying to attack, I have failed to notice my dependence on that ideology and so I am doubly 'immersed in the empiricist problematic' (p. 274). The authors correct this shortcoming, they offer a new view of human knowledge, they restore Law and Order and replace a merely verbal performance by the true radicalism of those in touch with social reality. Seen 'at least' from their advanced point of view, my book 'has little or no *importance* as a contribution to understanding the nature of the sciences and 'even less' to ethical/political theory. But it has what may be called a *significance* as an index of the contemporary crisis of empiricism and liberalism' (p. 249, cf. pp. 337–8). 'Ethico-politically', they write towards the end of their essay, '[my] position illustrates the threadbareness of contemporary liberalism,

---

[1] 'Feyerabend's Discourse Against Method: a Marxist Critique', by J. Curthoys and W. Suchting, *Inquiry*, Summary 1977. My reply was published in the same issue. Numbers in brackets refer to pages of the review. Numbers in brackets preceded by *AM* refer to pages of my book. CS refers to the authors of the review.

becoming increasingly disordered and politically ambiguous as it becomes
more and more irrelevant to the current scene of deepening capitalist
crisis and the increasingly powerful responses of the oppressed to that
crisis. In such a situation, and in the hands of class-peripheral, parasitic
intellectuals, liberalism becomes stripped to its bare constituent atom, the
single individual, posturing about in despair or self-congratulation (or
different mixtures of both), often spouting *enfant-terrible*-ish pseudo-
radical rhetoric the while' (p. 338), after which they dismiss the startled
(or slightly amused) reader with a pocket history of modern empiricism
from the Vienna Circle down to the author of *AM*.

One must admit, our two southern rhapsodists have studied the
Marxist vocabulary well. They are not too original[2] and there are certainly better stylists even among contemporary Marxists. Still, they know
the right words and they know how to put them together. But Marxism
is not just an inventory of phrases, it is a *philosophy* and it demands from
its practitioners a little more than a pure heart, strong lungs, and a good
memory. It demands from them the ability to recognize an opponent, to
separate him from other, though related opponents, it demands a nose for
differences that might seem insignificant when compared with the 'great
questions of the time',[3] which in turn demands an ability to *read* and to
understand what is said. Here, I am sorry to say, our Marxist friends fail
miserably. I am no longer surprised when finding that they ascribe to me
views I never even considered holding, for this is a familiar custom among
reviewers.[4] I even admire them for the verve with which they raise the
custom to a new level of excellence: they don't just misread the *book*, they
misread *their own review*; they quote me extensively and then, a few
pages after the quotation (or before it) take me to task for saying what I
don't say or not saying what I do say in the quotation. No doubt they first
made up their minds that I was a no-good, big-mouthed liberal-empiricist
bum and then adapted their mental reactions to this image. But I am
astonished to find two philosophers so unfamiliar with elementary principles of the art of argumentation. I am really embarrassed to bring the
matter up and hope I shall be forgiven for beginning my reply with a short

[2] For the passage just quoted they had to borrow from the arch-reactionary, Ernest
Gellner, whom they name without the slightest hint of embarrassment.
[3] 'The divergences between the Churchills and the Lloyd Georges ... are quite minor and
unimportant from the standpoint of ... abstract communism ... that has not yet matured to
the stage of practical ... action. But from the standpoint of this practical action ... these
differences are very, very important.' Lenin, *'Left Wing' Communism, an Infantile Disorder*,
Foreign Language Publishing House, Peking 1965, p. 99.
[4] For details cf. Chapter 4.

lesson in baby logic. The bored reader is advised to turn at once to Section 2 where the argument itself starts.

1. A Guide for the Perplexed

An important rule of argumentation is that an argument does not reveal the 'true beliefs' of its author. *An argument is not a confession*, it is an instrument designed to make an opponent change his mind. The existence of arguments of a certain type in a book may permit the reader to infer what the author regards as effective persuasion, it does not permit him to infer what the author thinks is true. It is interesting though somewhat surprising to see to what extent modern critics share the Puritan desire to 'talk straight',[5] i.e. always to speak the truth, and how often they misread more complex forms of argumentation (*argumentum ad hominem, reductio ad absurdum*) in this light. To aid them, here is a list of relevant rules with explanations and examples taken from the review.[6]

Basic Rule: If an argument uses a premise, it does not follow that the author accepts the premise, claims to have reasons for it, regards it as plausible. He may deny the premise but still use it because his opponent accepts it and, accepting it, can be led in a desired direction. If the premise is used to argue for a rule, or a fact, or a principle violently opposed by those holding it, then we speak of a *reduction ad absurdum* (in the wider sense).

Example: CS note that I draw sceptical conclusions from 'empiricism'.[7] They infer that I am an empiricist.[8] The basic rule shows that the inference is invalid:[9] the authors never took seriously my warning (quoted by them

---

[5] Not all Puritans were that narrow-minded and some of them cultivated the art of rhetoric without qualms.

[6] The rules were introduced by the sophists. They were systematized by Aristotle in his *Topics*. Only few modern readers seem to know how to apply them.

[7] In the sense of CS (pp. 262, 266, 290, etc.): separate existence of subject and object, need to establish a correlation between the two, Tarski's theory of truth, theory-ladenness of observations, Methodism, i.e. the belief in universal and stable rules that are imposed upon knowledge from outside and guarantee that it is 'scientific'. Minor features will be mentioned in the course of the reply.

[8] 'Feyerabend is thoroughly empiricist' (p. 267); 'genuinely empiricist' (p. 266); 'we have identified Feyerabend's epistemological position as empiricist' (p. 332); and so on and so forth.

[9] And *AM* shows that the conclusion is false: I do not accept theory-ladenness (*AM*, pp. 162f. and below in the present note), I regard the rules of Methodism only as a special case of the restrictions that affect a scientist (*AM*, p. 187, n. 15, quoted by CS, p. 253), I regard subject–object accounts of knowledge as specially problematic attempts to understand

in extenso) that I intend to play 'the game of [empiricist] Reason in order to undercut the authority of Reason' (p. 256; *AM*, p. 33). They are correct when saying that I am 'in the empiricist problematic' (pp. 274, 290) and that within that problematic there is only the alternative between Methodism and Scepticism (p. 290). They are mistaken when thinking that I *accept* the problematic and the alternative, or that I am *subjected* to them. Quite the contrary, I *use* them to turn an attack upon Methodism into an argument for scepticism and thereby into a *reductio ad absurdum* of critical rationalism (which is a version of empiricism in the sense of CS).[10]

That the authors are unaware (ignorant) of the nature of a *reductio ad absurdum* (or an *ad hominem* argument) becomes evident from their remark that 'Feyerabend's argument against Methodism fails because it is produced within Methodism's own empiricist problematic' (p. 332) – i.e. it fails because it is a *reductio ad absurdum* of Methodism. Almost all the more specific objections contained in the review as well as the Grand Indictment (PKF still an empiricist and he doesn't even know it!) rest on this unawareness (ignorance) and they collapse with its removal. This takes care of about three quarters of the review.[11]

The basic rule has simple *corollaries*.

---

our role in the world (*AM*, Chapter 17), I reject methodologies that impose rules from the outside and recommend a functional study of scientific procedure (*AM*, pp. 251ff. – subsections 2, 5, 6, 7; p. 260) as well as a 'cosmological criticism' of methodologies (*AM*, p. 206) instead – and so on and so forth.

Let us look a little more closely at the matter of theory-ladenness, which plays a large role in CS's picture of me. Theory-ladenness means that there is a theoretical load and something non-theoretical that carries the load contained in every observation statement. I have opposed this thesis in all my writings starting with my dissertation (1951) down to the last (the paperback) edition of *AM*. In *Proc. Arist. Soc.* 1958 I proposed to interpret observation statements in theoretical terms *exclusively*, in 'Das Problem der Existenz Theoretischer Entitäten', *Kraft Festschrift* (Springer-Verlag, Vienna 1960) I showed that the idea of theory-ladenness leads to paradoxical consequences, in 'Explanation, Reduction and Empiricism', which occurs in CS's bibliography, I tried to explain what one usually calls the 'observational core' of an observation statement *psychologically*, i.e. without reference to a division in the *content of the statement*, or in the *nature of the object* to which it refers, in 'Science without Experience' (*Journal of Philosophy*, Vol. LXVI [1969], reprinted in the *article*, 'Against Method', which also appears in CS's bibliography), I further strengthened my arguments against theory-ladenness. This short note which brought upon me the wrath of Ayn Rand (cf. the open letter to all U.S. philosophers of 3 April 1970 and her article in the *Objectivist* of March 1970 which has many similarities with the review by our Marxist friends) is summarized in *AM*, pp. 262f. CS must have very much wanted to turn me into an empiricist to overlook such clear evidence to the contrary.

[10] Thus it is true that I 'cele[brate] . . . the only alternative . . . empiricism has in face of the failure of Methodism' (p. 278), viz. scepticism. This is not because I myself regard scepticism as the only alternative to Methodism but because my opponents do and because the failure of Methodism means the end of *their* enterprise.

[11] 'Our critique of *Against Method*', write CS, 'is centred upon the claim that its major

First Corollary: If my opponent accepts historical facts and interpretations of historical events that can be used against him, then these facts can be used against him without any attempt to establish their validity.

Example: The authors chide me for using Galileo against Methodism without having shown that and why he is superior to Ptolemy–Aristotle. No such demonstration is needed, for Galileo is one of the heroes of empiricism (critical rationalism).

The authors also comment on the 'inadequacy of [my] theoretical justification of counterinduction' (pp. 262, 265). They fail to see that a theoretical justification is neither needed nor attempted. Using principles and historical facts accepted by the empiricist one finds that the heroes of science practised counterinduction.[12] That suffices to create a difficulty for the empiricist. I have no wish to go further, and I say so quite explicitly.[13]

Second Corollary: In an argument against an opponent an author can use assumptions and procedures he has shown to be unacceptable elsewhere provided they are accepted by the opponent.

Example: In the Brownian motion example I assert that a multiplicity of theories will produce more 'facts'. On the other hand I argue that there is 'no criterion for "facts"'. 'In the absence of such a criterion', write CS, 'there is simply no basis' for the assertion (p. 263). But the Brownian motion example is addressed to the empiricist who claims to have a criterion of factuality. I invite him to use this criterion in conjunction with my analysis and I foresee that he will be persuaded to become a pluralist (or, if he prefers monism, to take facts less seriously).[14]

Third Corollary: Having used part of a general view E to produce a result repulsive to those who accept E one may describe the result in terms of E thus stressing its distressing (for the defenders of E) aspects. If the result concerns a situation which the defenders of E hold in high

---

themes are all generated by an *unconscious* "empiricism"' (p. 266, my emphasis). So they *did* notice that I don't defend empiricism. But not understanding indirect arguments (where one uses a position in order to undermine it) and being entangled in their own ideology, the only avenue left open to them was to say that I defend empiricism 'unconsciously'.

[12] 'There is no doubt,' write CS, 'that . . . Galileo did proceed counterinductively' (p. 264).

[13] Cf. *AM*, pp. 32f.: 'One might get the impression that I recommend a new methodology which replaces induction by counterinduction . . . This impression would certainly be mistaken. My intention is not . . .' – the remainder being quoted by CS, p. 256. Cf. also my note in the *British Journal for the Philosophy of Science*, Vol. 27 (1976), pp. 384f.

[14] For general remarks on this pattern of argumentation cf. *British Journal for the Philosophy of Science*, op. cit., Section 1. The remarks in the text also defuse CSs' observation that I '[avoid] drawing the full sceptical consequences of the theory-ladenness thesis' (p. 262).

regard then we obtain paradoxical-sounding formulations (for the defenders of *E*, that is).

Example: I speak of Galileo's 'trickery', of his 'propaganda moves', I call his observation language 'metaphysical', his procedure 'irrational', I speak of the 'subjective', or 'irrational' nature of theory change (in Chapter 17) because this is how an empiricist who has followed and accepted my description of certain episodes in the history of ideas will have to describe the situation, and not because I regard the description as the last word on the matter.[15] Putting it briefly, I argue as follows:

'Look, my dear friends and critical rationalists! Here are some events in the history of science which for you are most important steps in the development of a new and rational world view. Yet, using your own conceptual toolbox, all you can say is that they are irrational, subjective,

---

[15] Example: in Chapter 12 of *AM* I briefly discuss a philosophy that makes sense of Galileo's procedure or, to use less neutral terms, makes it 'rational'. This philosophy (a) views ideas in their development and not as stable and unchanging entities, thus defusing the Methodists' objections to equivocation; it (b) keeps different strands of argumentation separate, thus preventing friction between them; it (c) takes into consideration the phase differences between new ideas and material manifestations of the old ideas on the one side, new ideas and accepted standards on the other, thus removing unsuitable criticism; and it (d) describes briefly the function of social forces. These social forces, I say, 'do not produce any new arguments' (*AM*, p. 154), but they reduce the frictions described in (b) and (c) and thus facilitate the growth of new theoretical objects (in CS this *distinction* between arguments which are parts of theoretical objects and conditions furthering the growth of the objects themselves becomes a 'fail[ure] to distinguish' between 'the necessity of . . . social factors for the development of theory and the fact that they do not intervene *within the production of the theory itself*' [p. 297]).

With incommensurability the situation is exactly the same. Cf. n. 38 of the present reply. In my analysis of Galileo I tried to achieve two things. (1) I tried to show the limits of empiricism (in the sense of CS), and (2) I tried to give an account of Galileo's procedure. As far as I can see, the major students of Galileo's methods (including Clavelin) tried to find a key, a particular way of doing research, that would unlock the 'secret of Galileo'. Occasionally the key was thought to be philosophical: Galileo the *empiricist*, Galileo the *Platonist*. In Clavelin we find the attempt to make *simplicity* a guiding notion. All these attempts overlook the 'opportunistic' character of Galileo's research (for 'opportunistic' cf. the Einstein volume of the Library of Living Philosophers [ed. by P. A. Schilpp]; Evanston, Illinois 1951, pp. 683f.). They overlook that Galileo *changes* his method whenever he thinks nature has changed its procedures, or whenever he runs into obstacles created by human psychology and physiology (cf. his account of irradiation). Nor does he avoid rhetoric. He is one of the few scientists (and in this way very different from Descartes, for example) who realizes that it is not enough to 'strive for the truth', but that the way to the truth must be made *visible* for man and that purely logical procedures must therefore be used side-by-side with rhetoric. Seen from an empiricist point of view much of what Galileo does is deception; seen from a more enlightened point of view it shows a marvellous insight into the complexities of knowledge. The only person who to my mind has given an account of how these different elements interpenetrate in Galileo to form arguments is Maurice A. Finocchiaro in his as yet unpublished essay on the logic of Galileo's *Dialogue*.

etc. etc.'[16] All this is beyond McEvoy[17] who thinks that I *subscribe* to the views that provide me with my terminological ammunition[18] and CS who sternly point out that 'initial "empirical inadequacy" is precisely what is to be expected in this [Galileo's] situation' (p. 298) – as if I thought otherwise.[19]

Fourth Corollary: If an author has a theory consisting of a part A which he regards as necessary protection against anarchy and another part B that may guarantee such protection if interpreted in a manner different from his, then it is legitimate, after removal of A, to tell him that nothing now separates him from anarchy, for he himself would describe the situation in this manner.

Example: Lakatos's theory of research programmes contains standards and it contains an account of theory production. According to Lakatos it is the former and the former only that make an enterprise rational and protect it from anarchy (Lakatos still believes in the distinction between a context of discovery and a context of justification). Hence, showing that he has no reason for accepting the standards and that the accepted standards have no force means showing the irrationality of his enterprise.[20]

Further corollaries can be easily produced and applied to CSs' criticism of AM. The result is always the same – most of this criticism is simply mistaken. I am not an empiricist, not even in the wide sense defined by CS, I don't accept Methodism, I don't accept the theory-ladenness thesis

[16] One of the few reviewers to notice this feature of my arguments is G. N. Cantor, *Journal for the History of Science*, Vol. 14 (1976), p. 273: 'This form of discourse may be appropriate for Feyerabend in his duel with Lakatos', and yet he continues, 'but', overlooking that AM is almost entirely such a duel (cf. the prefatory note, AM, p. 7 quoted, but not taken seriously by CS).

[17] CS quote McEvoy as authority for their statement that 'Feyerabend's historical examples are quite peripheral to his case' (n. 22), never realizing that McEvoy bases this judgement on earlier work of mine where I still defend the primacy of method over history (cf. McEvoy's quotation from me on page 51 of his essay) and thus revoking their own decision (n. 3) to develop the case from AM exclusively.

[18] The historical examples, writes McEvoy (*Philosophy of Science*, Vol. 42 [1975], p. 65) 'evince in the mind of *The Popperian* Feyerabend . . .' (my emphasis). Cf. also ibid., p. 64 where I am characterized as a 'sceptical fallibilist clutching at (*rationalist*) straws'.

[19] Cf. the account of 'backward movements' in AM, pp. 152, 158 as well as the following passage from *Criticism and the Growth of Knowledge* (ed. by I. Lakatos and A. Musgrave, Cambridge University Press, London/Boston 1970, p. 204): 'it would be a complete surprise and even cause for suspicion if all the available evidence should turn out to support a single theory, even if this theory should happen to be true.'

[20] Like critical rationalists CS point out that the standards are not completely impotent, but are strong enough to *judge* a situation as either advancing or degenerating. But what is the use of a point of view where a thief can steal as much as he wants and is praised as an honest man by the police and the common folk alike if only he tells everyone that he is a thief?

and subject–object correspondence as a condition of correctness, I argue against all these things 'right before their eyes' (cf. p. 287) and much of my 'pseudo-radical rhetoric' is empiricist (positivist) rhetoric turned against empiricism. How do all these mistakes arise? What is the reason for this almost superhuman blindness? There are some minor reasons, well known from standard scholarly exchange: an inability to read plain English, a tendency to lose the drift of an argument when the argument exceeds a certain degree of complexity, a readiness to merge a view with more familiar views that are vaguely similar to it. But in the case of CS these minor reasons are organized and held together by a fantasy that blots out all contrary evidence. And this is why I mention the matter. I don't mention it to 'prove my innocence', I mention it to show the power of antediluvian stereotypes over clear thought: the authors recognize that I am in favour of 'freedom' and that I have not much respect for institutions, however 'rational'. They do not examine what I mean by freedom, what I think about it and how I try to achieve it, they make no attempt to find out to what extent I agree with existing views and where I differ from them, no, they consult their party line, entry 'freedom', and come up with the master equation 'freedom = unconditional or absolute freedom = liberalism' (p. 249 and *passim*). Liberalism, of course, was once closely connected with empiricism (and still is, to a certain extent) and so they infer that I must be an empiricist as well. Note that all the inferences have so far occurred inside the stereotype, they have not yet touched 'reality'. But 'reality', i.e. my book, is easily dealt with. I often use empiricism in my arguments and I criticize ideas from an empiricist point of view. Connecting this 'fact' with the stereotype via some careless reading (cf. what was said above about minor reasons for the mistakes made) the authors can now boast about having independent evidence for their interpretation. And so the analysis proceeds without being troubled by contrary evidence, lending further strength to the stereotype.

Finally, two questions that might arise from my reply as well as answers to these questions.

Question Number One: If I am not an empiricist, then why do I argue in a manner that is liable to mislead faithful, hardworking though perhaps not excessively bright Marxists?[21]

Answer: I take seriously Chairman Mao's 'third indictment against

---

[21] 'Outsiders', i.e. laymen, scientists and just plain folk, do not seem to have been misled in this manner though they have another complaint: they just don't want to live without Methodism. 'What shall we do?', they constantly ask me, as if it were not *their* task to look for the methods they need for their work.

stereotyped Party Writing' which is that 'it shoots at random, without considering the audience' and I also take seriously his advice (which is old hat with rhetoricians but has been adapted by him to a new situation) that writers 'who really want to do propaganda must consider their audience and bear in mind those who will read their articles and slogans or listen to their speeches and talks'.[22]

Whom do I consider to be my audience?

First, all empiricists whether of the verificationist (probabilist) or the falsificationist persuasion, and this means practically all Anglo-American philosophers of science. Also many scientists who have strong convictions as to how science should proceed.[23] And a large part of the general public that is hypnotized by science and by Methodism.

Secondly and mainly, Imre Lakatos.[24] When calling things 'irrational', 'arbitrary', 'subjective' I am using words Lakatos himself used with gusto against opponents in sociology, quantum mechanics, and the philosophy of science. I attempt to show that as things stand within the methodology of research programmes he must apply these words to Great Science as well. In this I seem to have succeeded for in his last year Lakatos ceased to combat scepticism by the methodology of research programmes and started appealing to common sense instead.[25]

Question Number Two: Is the anarchism of the book merely a polemical instrument or do I want my readers to take it seriously as a positive philosophy?

Answer: Both the former and the latter! Anarchism is used in *AM* in two ways, viz. (1) as a polemical weapon, and (2) as a positive philosophy.[26] Accordingly there are two types of argument, viz. (a) arguments that establish its polemical function and (b) stronger arguments that show that it has intrinsic importance as well. I don't accept the premises of the first arguments, but I do accept the premises of the second. It is interesting to see how the situation looks to CS. Being unaware of the Basic Rule and its Corollaries but very much aware of the liberal-empiricist stereotype,

---

[22] 'Oppose stereotyped Party Writing', *Selected Works of Mao Tse-Tung*, Vol. III, Foreign Language Publishing House, Peking 1965, pp. 58f. Mao speaks of 'communists', not of writers in general.

[23] That these convictions often conflict with what they are doing as scientists is realized by CS: p. 299, item 4.

[24] Cf. the introductory note to *AM*.

[25] Cf. 'Imre Lakatos', *British Journal for the Philosophy of Science*, Vol. 26 (1975), p. 17 as well as Appendix 4 of *AM*.

[26] Even here I don't defend anarchism as an 'eternal philosophy' but as a 'medicine' (*AM*, p. 17) that may have to be withdrawn when the conditions change (*AM*, p. 22). The restriction remained unnoticed by CS (p. 252).

they read (1) as (2) and (b) as (a) and so they can criticize me for using the weak arguments, (a), for establishing the stronger case, (2). They do notice that (a) are not the only arguments in the book and they quote two arguments of type (b), but they call them 'subsidiary' (n. 16) and decide 'not . . . to discuss' them 'preferring to concentrate on the central arguments' (n. 11) – which are precisely the arguments whose premises I do *not* accept. So we have here the amusing spectacle of two reviewers misreading the polemical use of a position as a direct defence of it and taking me to task for not giving sufficient support, while regarding the stronger arguments that provide the support as 'subsidiary' and 'not central'. Let us now take a closer look at this matter of epistemological anarchodadaism.

2. Discourse on Method

In order to explain the polemical use of epistemological anarchism (EA for short) it is convenient to discuss the following methodological positions (some of which are Methodist in the sense of CS):

(A) Old fashioned, or naive rationalism (Descartes, Kant, Popper, Lakatos; ancestor: the philosophy behind the apodictic laws of *Exodus*).

(B) Context-dependent rationalism (some Marxists, many anthropologists and anthropological relativists; ancestor: the philosophy behind the case laws of *Exodus*, which is older than the apodictic philosophy and comes from Mesopotamia;[27] it can also be found in prehomeric Greece and in the China of the oracle bones).[28]

(C) Simple-minded anarchism (some ecstatic religions and various forms of political anarchism).

(D) My own view (ancestors: Kierkegaard's *Concluding Unscientific Postscript* and Marx's comments on Hegel's *Philosophy of Law*).

According to (A) it is rational (proper, in accordance with the will of the gods) to do certain things *come what may* (it is rational to prefer the more probable hypothesis, to avoid *ad hoc* hypotheses, self-inconsistent theories, degenerating research programmes). Rationality is universal, stable, independent of content and context, and it gives rise to equally

---

[27] For the distinction between apodictic laws and case laws and their historical ancestors cf. W. F. Albright, *Yahweh and the Gods of Canaan*, Anchor Books, Doubleday & Co., New York 1968, Chapters 2 and 4.

[28] For details cf. my *Rationalism and the Rise of Western Science* forthcoming.

universal rules and standards. Some reviewers, including CS, have classified me as an old-fashioned rationalist in this sense with the proviso that I try to replace the traditional requirements of old-fashioned rationalism with the more 'revolutionary' requirements of proliferation, counterinduction and so on. This matter has already been dealt with.[29]

According to (B) rationality is not universal, but there are universally valid conditional statements asserting what is rational in what conditions and there are corresponding conditional rules. This also has been regarded as the 'essence of my position'. Now while it is true that I frequently comment on the need to take the context into account, I would not introduce it in the way recommended by context-dependent rationalists. For me the rules of context-dependent rationalism are just as limited as the rules of old-fashioned rationalism.

The limitation of all rules is recognized by (C). (C) says (a) that both absolute rules and conditional rules have their limits so that even a relativized rationality, when followed to the letter, may prevent us from reaching what we want to achieve, and it infers (b) that all methodological rules are worthless and should be given up. Some reviewers thought that Ca and Cb taken together expressed my position, overlooking the many passages where I show how certain procedures *aided* scientists in their research. For in my studies of Galileo, of Brownian motion, of the Presocratics I not only try to show the *failure* of traditional methodologies (of the Methodist type), I also try to show *what procedures did actually succeed* in these cases, and why they succeeded. Thus I agree with Ca, but I do not agree with Cb. I argue that all rules have their limits, I do not argue that we should proceed without rules. I argue for a contextual account, but the contextual rules are not to *replace* the absolute rules, they are to *supplement* them. In my polemics I neither want to eliminate rules, nor do I want to show their worthlessness. My intention is, rather, to expand the inventory of rules and also to suggest a new use for all of them. It is this *use* that characterizes my position and not any particular rule-*content*. I shall now say a few words about this matter.

Absolutists and relativists of type (B) obtain their rules partly from tradition, partly from abstract considerations concerning the 'nature of knowledge', partly from an analysis of more specific, but still absolutely conceived conditions (the conditions of man, in the case of Kant's absolutism). They then assume that each individual action, each individual piece of research must be subjected to the rules they have found.

---

[29] Cf. above, n. 13 and text.

The rules (standards) determine the structure of research in advance, they guarantee its objectivity, they guarantee that we are dealing with rational, scientific action. As opposed to this *I regard every action and every piece of research both as a potential instance of the application of rules and as a test case*: we may permit a rule to guide our research, or the kinds of actions we are interested in, we may permit it to exclude some actions, to mould others and on the whole to preside like a tyrant over our activities, but we may also permit our research and our activities to suspend the rule or to regard it as inapplicable even though all the known conditions demand its application. In considering the latter possibility[30] we assume that *research has a dynamics of its own*, that it can proceed in the absence of clearly formulated rules and that research so conducted is substantial enough to gain attention from the defenders of the *status quo* and orderly enough to serve as a source for new and as yet unknown procedures. This assumption is not as unrealistic as it sounds. It is taken for granted by those who defend the distinction between a context of discovery (which proceeds without any guidance by rules) and a context of justification (where rules are applied); it can be supported by pointing to the inventiveness of the human mind which responds to unforeseen problems with all sorts of ideas as well as to the inner dynamics of forms of life which add their own transindividual structure to the achievements of individual thinkers. Nobody can anticipate the shape of the products that arise in this way, nobody can say which rules and standards will be suspended and which will remain in force and nobody can therefore guarantee the permanence of rules of type (A) and (B). This, in essence, is my objection to methodism. The objection is strengthened by case studies which show *how* rules were suspended and what procedures took their place.[31] No system of rules and standards is ever safe and the scientist who proceeds into the unknown may violate any such system, however 'rational'. This is the polemical meaning of the phrase 'anything goes'.

Note that the argument rests on the assumption that forms of life which are permitted to suspend Methodist rules are taken seriously by the Methodists. The argument does not provide a way of judging their merits, it accepts the judgement ('Galileo a great scientist!') of the oppo-

---

[30] What follows is an elaboration of ideas found in Chapters 2 and 17 of *AM*.

[31] The new procedures are not always 'known' to their users, they are not explicitly formulated and then followed; no, they are introduced in the guise of respectable rules though much twisting and turning has to be done to adapt the idol to reality. Newton is an excellent example.

nents. Note also that rules and standards are not abolished – one does not enter research without any Methodist equipment – but are used tentatively and changed when the results are not as expected. These changes do not prove that there are more general rules which decide when specific rules can be used and when they have to be suspended for individuals, when behaving in an orderly manner, both constitute rules and follow them. Note, finally, how this argument for scepticism differs from the argument CS ascribe to me:[32] it does not enlarge on the difficulties of the subject–object distinction, it considers the precarious situation of rules *vis-à-vis* the development of existing practices[33] and the appearance of new ones and thus belongs to the 'Third World' entirely – if one likes to express simple matters in pompous language.[34] This concludes my

[32] According to CS the 'dramatis personae' of 'traditional empiricism' are: 'the knowing subject face to face with a real object . . . [t]hen the picture of this subject characterized by beliefs' etc., etc. (p. 339). The remarks in the last section as well as the brief account of scepticism given just now show that these characters never appear in *AM*, not even in the polemical arguments.

[33] One should note that the argument does not compare rules with a practice that is independent of them – it does not reject rules, because they do not fit 'history'. The argument is, rather, that the critical rules, *when introduced into existing practices*, i.e. when allowed to interfere with them will disturb them in a most undesirable fashion. They will create results the defender of the rules is not likely to approve of.

[34] CS take great pains to show that Popper (whom they liken to Faust – really kids!) is a part-time Marxist and they criticize me for the 'sin' (the SIN?) of overlooking Popper's more reasonable (i.e. more Marxist) ideas. I don't quite understand how they can say that, for in my review of *Objective Knowledge* (*Inquiry*, Vol. 17 [1974], No. 4), to which they refer but apparently without having read it, I deal with these 'reasonable' ideas exclusively. I point out that they don't belong to Popper but go back at least as far as Aristotle, who expressed them in a simpler and less technical fashion: Aristotle started the study of the history of ideas because he believed in a 'third world' created by men. I discuss some of Popper's arguments and show that they are either simple declarations of faith, or bad rhetoric. I criticize his tendency to solve problems of reduction by a swift ascent to higher realms of being (scientific research and not philosophical manoeuvres must decide whether psychological phenomena are 'reducible' to material processes). And, finally, I show that Popper constantly mixes up the distinction between relative causal autonomy and ontological difference. For him the laws of arithmetic deal with entities which are not material, while I assert that they are autonomous laws *of matter* which are causally *but not ontologically* differentiated from other laws of matter. Now before CS speak of 'reductionism' (a favourite charge raised by obscurantists against those who take the trouble to study the laws of matter in greater detail) they should read my arguments against Popper's interpretation of arithmetical laws and my own positive suggestions (which are not at all different from suggestions found in Engels and, in our days, in Hollitscher). Let them raise the charge of reductionism in *connection with such a criticism* and not in empty space and they will see that the matter is not at all simple and that Popper seemed a part-time materialist only because none of the present-day materialists has the perseverance to read all his 'arguments' in detail and the intelligence to see their flaws. Besides, all they care about are utterances which vaguely agree with what they think is Marxist party-doctrine, they do not care about analysis and the improvement of that doctrine. That applies to CS, Althusser, and all the rest.

account of the *polemical use of scepticism* in *AM*.

Turning now to CSs' *argument against scepticism* we first note that it is not an *argument*, but simply a (long-winded) *rejection* of scepticism. CS formulate a view, Marxism, which in their opinion excludes scepticism. This is nothing extraordinary, for there are many such views. For example, scepticism does not arise in a world without humans, and it does not arise in a world with obedient humans who never stray from the *status quo*, not even in their thoughts. And then there are other views which provide ample support *for* scepticism. CS never argue for the view *they* adopt. They simply say that they will criticize scepticism 'from a Marxist point of view'. The most they can show, therefore, is that scepticism is not good Marxism, they cannot show that it is incorrect.

The remark can be generalized. Every criticism CS make of me consists in showing a conflict with Marxism. Hence, even if their criticism were to hit its target (which, we have seen, it does not) it would at most amount to showing that I am not a Marxist and not that I am not right.

CS attempt to show that Marxism excludes scepticism. The attempt does not work. According to CS scepticism is removed by a practice that not only *determines the actions* of the theoretical labourer but also *produces the objects* of his science. A scepticism that derives its force from the difficulty of correlating two distinct and separate entities, an isolated 'absolutely free' subject (p. 334) and an object that is 'totally other' (p. 289) now seems to lose a great deal of its plausibility. The apparently unreachable object not only can be reached, it can even be produced.[35]

Now it would be very naive to assume that scepticism is thereby done away with. Production often misses the real object, just as correlation does. The practice of research itself has proved many instances of production to be illusory. At the height of the witchcraft persecution demons *were produced*, ordered to perform complex tasks, they *performed* these *tasks* not only in the imagination of contemporaries but according to the practices of which these contemporaries were the 'bearers' (p. 271) (much of witchcraft was spiritual engineering). With the aid of careful and rather elaborate preparations mystics could rise beyond the Seventh Heaven and see God in all His Splendour and others could turn into animals and back again. Phlogiston, ether, N rays, magnetic monopoles, inheritance of acquired properties were all produced at some time and denied existence at others. Of course, we can say today (or, remembering earlier errors, we *think* we can say today) what was really produced, but the point

---

[35] The argument from practice and the historical view based on it is due to Althusser.

is that while all the criteria of production pointed to one object, another object was actually present. In this, production is not at all different from correlation. Moreover, it not only turned out that 'production' had failed to get hold of the real object and had 'produced' a chimera instead, it also turned out, and on the basis of the very same arguments, that the real object contained non-producible parts and could therefore not be entirely grasped by a science in the sense of CS: the development from Aristotle to Newton led from a space that was producible and whose parts were producible (cf. Aristotle's theory of the continuum in Book II of the *Physics*) to an entity no part of which could be changed or produced by natural means. Similarly, producible forces were replaced by the non-producible force of gravitation.[36] And more recently a producible (via the actions of the ether) speed of light and producible features of elementary particles were replaced by absolute constants and removed from the domain of CS-science. Following scientific practice ('scientific' now taken in the usual sense) we therefore learned that the world contains producible entities side by side with isolated entities (entities which have effects but cannot be produced, neither by nature, nor by science),[37] that a science in the sense of CS is not complete and that production may be illusory. This dissolves CSs' master-'argument' against scepticism.

So far the master-argument has turned out to be defective in two ways: it does not offer any arguments for the view that is supposed to exclude scepticism, viz. the production view of knowledge, and scepticism is not excluded by this unsupported view. In addition the argument was never directed against scepticism as a whole but only against a particular version of it. Other versions still remain unconsidered.

These other versions can be developed independently of Methodism and the subject–object ideology and they lead to a strong presumption that scepticism and its generalization, epistemological anarchism, may after all be an acceptable account of the nature of knowledge.

To start this new argument let us remember that usually there is not just one practice to deal with a given real object, but that there are many. In medicine we have a Western 'scientific' approach (which arose from the extension of the problematic of 17th century science to the human

[36] It is amusing to see that Aristotle, whose philosophy is regarded as 'ideology' by CS (p. 298), gives a production-account of space, terrestrial objects, the continuum, even knowledge, while his successors introduced isolated entities: the transition from Aristotle to Newton contains transitions from scientific to non-scientific (in the sense of CS) accounts.

[37] 'Our central thesis', write CS (p. 296), 'is that Galileo's basic achievement . . . consisted in the fact that he laid the main part of the foundations of physics as a *science*.' Certainly not as a science in the sense of CS – cf. the preceding note.

organism) side by side with the medicine of the *Nei Ching* and tribal medicines. These practices are scientific in the sense that they either produce states of the organism or can say how such states are produced, they are successful for they heal and bring about other desirable transformations, and they all deal *with the same real object* but on the basis of radically different problematics. Each practice also determines the attitude of its practitioners. Of course each practice has faults and lacunae, but these faults are distributed in different ways among the different traditions. A 'neutral observer' who has not yet fallen under the spell of science and who judges the practices by the way they restore health would be hard put to make a choice. A 'peripheral' opponent of a particular practice, however, need not end up in no-man's-land. He may find a home in a different but equally acceptable tradition. *Exchanging practices need not reduce contact with reality.*

Secondly, let us remember that practices have their ups and downs, they start from insignificant beginnings and grow under the jealous glance of a well-endowed rival. These differences in theoretical sophistication do not always mirror differences in relation to the real object. Some practices may appear to be ahead of other practices through producing almost all the objects described, but the production may be a chimera while an as yet ineffectual rival may have a problematic that will eventually grasp reality in a more efficient manner. Result: *exchanging a powerful practice for a puny tradition need not reduce contact with objective reality* (though it will of course lead to a clash with social, or 'scientific' reality).

Thirdly, traditions are often removed by force and not because of 'autonomous' developments. Valuable knowledge disappeared under the pressure of external circumstances and not because it was found deficient: *exchanging a powerful tradition for a mere dream need not reduce contact with reality.* The world in which we live has many sides, many aspects, many potentialities. Doubters, peripheral thinkers, dreamers, mere fools *have a real chance* (and not just a logical possibility) of making discoveries which are beyond the reach of established traditions.

Let me repeat the ingredients of this argument so that it is clear what is assumed and what is asserted.

It is assumed that the real object permits practices with different problematics to have comparable achievements (i.e. to achieve a comparable balance of failure and success) and it is asserted that an individual who exchanges practices (who turns from science to the Tao) therefore need not lose touch with reality. It is assumed that practices may disappear for external reasons and asserted that an individual who leaves a powerful tradition and enters no-man's-land with but a few 'peripheral' fantasies to guide him need not lose touch with the real object (with God, with the material world, with Non-Being). He may and most likely will lose touch with the *social reality* that surrounds him and become an outsider. Losing touch with social reality he may also lose that 'sense of the real' that accompanies both production and sham production and makes it so difficult to distinguish the two. He may lose his sense of language and be reduced to stammering and inarticulation. Religious reformers and scientists alike had this feeling when venturing beyond the boundaries of the *status quo*. But they also felt strong objective guidance which means, to use CSs' terms, that there was a problematic hidden in their dreams and this problematic was about to unfold and to become a common malaise. Note to what extent socio-cosmological assumptions enter the argument: the two assumptions made above are generalizations from historical fact. Note also how carefully the theoretical (social) object is separated from the real object. It is this separation (taken in conjunction with the assumptions) which enables us to assert that peripheral thinking, dreaming, doubting need not lose contact with reality.[38]

Now how can our dreamer turn the *real chance*, i.e. his chance *given*

[38] A brief sketch of the argument is found in *AM*, pp. 20, 206, 293 (which, incidentally, are the arguments regarded as 'subsidiary' by CS – cf. the end of Section 1) and especially in the chapter on *incommensurability*. CS make a mess of this chapter, so let me give a brief account of what is contained in it.

First, the chapter contains a refutation of the empiricist idea that any two theories can be compared with respect to content or, that there are pairs of concepts, one belonging to one theory, one to the other, that can be brought into the relation inclusion/exclusion/overlap. To refute this idea I use the empiricist 'reconstruction' of theories, i.e. I use cross-sections of their development without regard to origin, problematic, details of growth (cf. CS, p. 323). I show that there are pairs of theories in this sense whose contents cannot be compared though they seem to deal with 'facts in the same domain'. I thereby show the limitations of all criteria involving content (even Lakatos uses a criterion of this kind). I do not infer that 'theories' (in the empiricist sense) are incomparable in other respects as well, nor do I assume that the empiricist account of theories is adequate.

As regards the first point I have considered a variety of criteria of comparison that do not involve content (for a survey dealing with my attempts from 1951 up to the present cf. Section VI of 'Changing Patterns of Reconstruction', *British Journal for the Philosophy of Science*, Vol. 28 [1977], forthcoming). These criteria were designed for empiricist 'reconstructions', but some of them have wider applications. As regards the second point I emphasize that a theory can 'never be fully separated from the historical background' (*AM*, p. 66), and give outlines of problematic and principles of construction (development) both for the Homeric universe (where the matter is rather difficult to ascertain) and for the Presocratics. I also explain the manner in which some theoretical objects are built up (physical objects and human: *AM*, p. 247; knowledge: *AM*, p. 246).

I also try to explain the vague notion, found in many art historians as well as in some followers of Wittgenstein (Hanson, for example), that we 'see' reality in terms of our concepts. The notion turns out to be true in special cases only and I try to determine these cases in greater detail. The very aim of the investigation shows that I don't 'conflate' (don't CS have a less ugly word to describe the process?) 'theory and experience' (p. 326). I (a) insist that theories are broader than experience (I wrote a note 'Science without Experience' – remember? Cf. n. 9 of this reply), and (b) go to a lot of trouble to show that it is only in special circumstances that theories shape experience in their image (*AM*, pp. 236ff. – CS quote a long excerpt from this argument but as usual don't understand it). Nor do I ever say that what is is the same as what is thought to be ('conflation' of theoretical object and real object). 'Realism' as defined in the incommensurability chapter does not mean that the real is *identified* with the theoretical object, it means that one *tries to understand* the real in theoretical terms rather than regarding it as 'given'. Such, at least, is *my* view on the relation between the real object, the theoretical object, and the experienced object. (At the end of Chapter 17 of the German edition of *AM* I present a different view; cf. also the end of this Note.)

Now *my view* is not always the *view of the cultures I examine*. Many cultures, some periods of science included, make no clear distinction between real object and theoretical object, and some that do make such a distinction draw it purely verbally, without theoretical significance (some versions of Kant's *Ding an sich*). Occasionally the distinction is made *and* leaves a trace in the theoretical object: human knowledge and human thought are inadequate for grasping God, and faith and revelation must come to the rescue. Now when discussing incommensurability I was not only interested in making life difficult for critical rationalists. I also wanted to understand the changes that take place when a new world view enters the scene. These changes can be examined in various ways. They can be examined 'from the outside', that is by looking at them from the point of view of a favoured philosophy (Marxism, in the case of CS). I don't deny that such an examination is possible and I don't deny that it may succeed in rationalizing all change (CS ascribe to me the belief in the 'impossibility of rationalizing all scientific change' [p. 331] – but I restrict incommensurability to *special types* of change and grant that 'external accounts' might succeed in rationalizing even these special cases [*AM*, p. 232]). But an external examination really does not interest me too much. I am not interested in how a particular event looks when projected on a later ideology, I am interested in how it looks 'from the inside', i.e. to the concerned parties. Can these parties make sense of the changes that are going on, can they subject them to what they regard as their own rationality, or are they forced to admit that they are part of a process they cannot master with the available forms of reason? This, incidentally, is also the question that arises at times of scientific revolution. The question is then not whether the whole conflagration will look reasonable five hundred years later, *but to what extent it can be made reasonable when it occurs and to what extent one must permit reason to be violated* ('reason' always meaning the form of reason that is accessible to the participants). Obviously such an examination is most important for every researcher. It prepares him for events which otherwise would catch him by surprise.

Now when examining traditions 'from the inside' we must adopt the ideas and the procedures of the participants, and we must try to reconstruct the world as it looked to them (their 'phenomenal world'). If the participants do not distinguish real object and theoretical object, then we must not draw this distinction either, and 'symptomatic readings' (pp. 328f.), which import external criteria, are out. This is the reason why I often disregard the distinction between real object and theoretical object and why I occasionally also disregard the distinction between perception and theoretical object. It is not I who 'conflates' what is to be kept apart, it is CS who conflate external accounts and internal accounts and introduce external criteria and distinctions where they have no room.

Finally, why should 'real object' and 'theoretical object' be separated? What are the reasons which our supercritical reviewers can give for this distinction? They give no reasons. They say the distinction occurs in Marxism – and that is that. Hence, even if their criticism were to hit its target, namely me (and we have seen on numerous occasions that it does not), this would merely show that I am not a Marxist and not that I am not right. The burden of *this* criticism, however, is easy to bear.

the structure of the real world into a *social chance*, i.e. how can he make his dreams *popular*?[39] By connecting parts of it with existing practices in such a manner that the popularity of the practices flows into the dream or else by telling his dream in a reasonable fashion that makes it merge with 'facts' and opinions of the time. It is fascinating to see how individuals and small groups falsified their dreams in the manner just described and then changed the (social) reality that provided them with the instruments of falsification.

So far I have given a thoroughly individualistic account of social change. This is not how matters look to some researchers. Viewing events with hindsight they often perceive an orderly progression of institutions, social conditions, and ideas without any major role for the individual. Our authors go further and sneer at the 'myth of creation' (p. 265) which regards individuals as starting-points of ideas. For them doubts, dreams, feelings of discontent are peripheral events that *accompany* an objective theoretical process but do not *guide* it. 'Individuals are . . . "bearers" of the relation of the theoretical production process in which they are engaged. Their actions, beliefs, etc., can be partially explained with reference to this process but not conversely' (p. 271). This view is the last obstacle in our path.

To remove the obstacle we admit that traditions, theories, problems obey laws of their own but add that their development is not governed by such laws exclusively. This can best be shown by a *computer analogy*.[40]

Computers can solve some problems, they cannot solve others, and they occasionally break down. Difficulties are dealt with either by repairing them, i.e. by changing their structure and/or their programme, though not drastically, or by replacing them. Replacing a computer means building a computer with a different programme and a different basic structure using partly new material, partly hardware from the computer that is to be replaced. Replacement is preceded by blueprints (metaphysics in the case of scientific traditions) and there are stages when the standard computer is run side-by-side with only partly functioning new computers which are built according to promising blueprints but often break down. Now let us consider a situation where we have a standard computer with clearly identifiable difficulties, alternative blueprints that relate the difficulties to basic features of its structure and programme,

---

[39] Popularity is not needed for cognition, but for the knowledge of cognition.

[40] This objectivizes knowledge, makes it relatively stable and independent of subjective feelings but without any ascent to a 'Third World'. Cf. n. 22 of my review of *Objective Knowledge*, *Inquiry*, Vol. 17 (1975).

a partially functioning realization of some blueprint that seems to work well in some areas, gives absolutely no response in others (though these other areas are dealt with by the standard computer), as well as stopgap measures for the difficulties of the standard computer (example: the standard computer = quantum field theory; the stopgap measures = renormalization; the alternative = hidden variable theories). Shall we continue using the standard computer, shall we abandon it and concentrate on the development of its rival, shall we try to develop still further rivals, shall we engage in all these things at the same time or what else shall we do? Here is a genuine problem. What is the solution?

The solution is not difficult if there exists a super-computer dealing with the relative merits of computers and issuing directives for their development. We leave the problem to her and she may give us an unambiguous answer. Theoretical development (the work of the super-computer), not personal decision decides what is going to happen. Of course the problem will eventually recur on the higher level and so on. Now I assume that *the historical process consists only of a finite number of 'levels' of this kind*. Results in decision theory show that not all problems of n-level computers can be solved by $n-i$ level computers, for any i. Hence we must concede that decisive historical developments are either chance events *or we must introduce the individual as a causal agent that changes aspects of traditions and brings about revolutions*. The latter interpretation means, of course, that dreams, feelings of doubt, peripheral 'subjective' ideas function as described above: they not only *reflect* social change, they can also *initiate* it. I shall adopt this interpretation. As a result we can now say that the world is so constituted that any attempt at subjective liberation, any attempt to develop one's own being has a real chance (and not merely a logical possibility) of contributing to social liberation and of improving our understanding of the real world[41]

[41] The assumption concerning the number of levels must of course be confirmed by a more detailed study of scientific revolutions (and other upheavals) than is available to us now. We must examine the *minds* of the participants, their memories, their habits, their convictions, their dreams. We must try to find connecting links between these elements and the theoretical activity in which they were engaged, and we must examine this theoretical activity in much greater detail (Stillman Drake's more recent analyses of some Galileo MSS is exemplary in the latter respect). Then comes the study of the role of the key individuals *in their profession*. What was their credibility, who listened to them, to what extent could they strain their reputation and still be taken seriously, to what extent *did* they strain their reputation? This gives us some connecting links between their theoretical activity and the *status quo* in their profession. (Here we may discover that it was not arguments that counted, but reputation pure and simple.) Then comes the examination of the role of the profession in society at large and the back effect, on the profession, of events outside the subject. Protestants such as Maestlin would regard the Pope's calendar reform with different eyes than

3. On Liberty

CS say that I assume absolute liberty. I say that absolute liberty is an abstraction not found in this world but that conditional liberty is possible, desirable, and should be sought. I also say that in our world conditional liberty is not just a luxury – though there is no reason why luxury should be avoided – but a way of coming to know new features of the world. It is extremely hard to achieve. To speak we have to internalize a language, to think we have to absorb further theoretical relations, to act and to succeed we must be familiar with the moods, the demands, the tricks of society and we must be able to react without reflection or reflection cannot even start. Our minds and our bodies are restricted in many ways. Nor does our education help us to reduce these restrictions. From our very childhood we are subjected to a process of socialization and enculturation (to use ugly words for an ugly procedure) compared with which the training of household pets, circus animals, police dogs is mere child's play. The noblest human endowments, the gift for friendship, trust, the need for companionship, the will to please that is to make *others* happy are misused and defiled in this process by teachers who have only a fraction of the talents, the inventiveness, the charm of their pupils. They are not entirely unaware of their shortcomings and they take revenge. For their one and only aim, their life's ambition is to reduce their wards to their own squalour and stupidity. Even intelligent and understanding teachers don't *protect* their pupils from being overwhelmed by the material they are supposed to absorb, they try to make the acquisition of this material *easier* thus putting liberty at a disadvantage from the very beginning. What is the result of this education? One meets it day in, day out, at our universities: servile non-entities who vainly try to identify the source of

---

would Catholics (today scientists in government protection agencies view evidence differently from scientists hired by business, and so come to different conclusions. This has been confirmed by a series of most interesting studies. Cf. 'Behind the Mask of Objective Science', in *The Sciences*, November/December 1976). The forces are small and remain undetected by those who enlarge the effects of reason to such an extent that nothing else can be seen, but small forces passed through amplifiers such as intelligent men with a good reputation may have large effects. None of these things are accessible to those who proceed from a pre-conceived idea of method, insist on a separation of internal and external affairs and similar restrictions. No wonder the debate continues and continues and continues.

Considering the assumption in the text: we can of course use the theory that emerges from a conflagration or some later theory for projecting an underlying structure on the conflagration itself and thus reveal 'inevitable developments'. But the point is that the rationalizing theory was not present at the time of the conflagration and could not lend structure to the actions of the participants. These actions were therefore genuine primary causes.

their misery and who spend the rest of their lives in the attempt to 'find themselves'. What they do find when proceeding with their studies is that lack of perspective is really 'responsibility of thought', that illiteracy is really 'professional competence', and that mental constipation is 'scholarship'. So elementary education joins hands with higher education to produce individuals who are extremely limited, unfree in their perspective though not at all in their determination to impose limitations on others under the name of knowledge. Have you ever seen a young cat face an unusual object? Its whole being is affected as if it were asked to become different from itself. Teachers and 'people who know' are affected in the same way, but they have learned to throw their discomfort back upon the world in the form of disapproval and contempt. But let us not unduly linger with the universities, for the situation is the same in churches, politics, the military establishment. Everywhere people without hope take hope away from those who still have it, encourage, badger, coax them to 'face reality' and thus make sure that the world will never lack the likes of them.

The mistreatment of minds is accompanied by the mistreatment of bodies. Medical science has by now turned into a business whose purpose is not to restore the *natural state* of the sick organism but to manufacture an *artificial state* in which undesirable elements no longer occur. It triumphs in areas of surgical intervention and is almost entirely helpless when faced with disturbances that involve the balance of the organism such as certain forms of cancer. The technological approach with its in-built distrust of nature, its conceited belief in the excellence of science, and its determination to remake man and nature in its own image naturally favours surgery even in minor cases which could easily be restored by other means. Thousands of women lost their breasts when they could have been healed by simple massage, diet, acupuncture, herbal treatment. There are various reasons why this fantastic incompetence of modern scientific medicine remains hidden from the public. First, modern medicine defines its own standards. A mutilated body that can barely drag itself along and has to be sustained by pills, injections, kidney machines, occasional follow-up operations is 'the best modern science can do for you'. A second reason is that the enormous amount of research being carried out and which, like Vietnam, always holds out the promise of a 'breakthrough', lends respectability to what otherwise would be nothing but systematic bungling. Thirdly, we must not forget the public's fascination with gadgets. The machinery used by modern medicine is often superfluous, any Chinese country doctor can diagnose much better from pulse, urine, texture of skin, self-report of patient – but who is nowadays going to prefer pure and unadulterated human ingenuity to technological display? Then there is television where the incomprehensible and marvellous professional lives of dedicated body-plumbers are combined with their comprehensible and not at all marvellous private lives, thus creating an awesome and most attractive mixture of public service and personal tragedy. Finally, and most importantly: modern scientific medicine *lacks the necessary outside checks*. We *have* the traditions that could expose its so-called productions. But these traditions are not permitted to work: killing in the scientific manner is legal while healing in the non-scientific manner is outlawed.[42] *This* is the reality (only a microscopic part of it!) which we have to address, this is the reality which CS enshrine in their idea of a practice, this is the world they defend by sneering at those who look for a more pleasant life. It is true that Marxism once went a different way and had different aims. But the vision of the founders has now become a doctrine, their insights have been buried in footnotes and the small group of humanitarians has turned into a swarm of intellectuals who criticize other intellectuals and are taken to task by still further intellectuals, a tearful line here and there replacing the humanitarianism that is absent from the whole enterprise.

In the face of these mind-killers and reason-mongers, in the face of these scientific mutilators of body and spirit I try to defend the liberty of the individual, his right to live as he sees fit, his right to adopt the tradition he reveres, his right to reject 'truth', 'responsibility', 'reason', 'science', 'social conditions' and all the other inventions of our intellectuals, and also his right to an education that does not turn him into a mournful ape, a 'bearer' of the *status quo*, but into a person who is able to make a choice and to base his whole life on it. I *defend* this right – but how can it be *realized*? And will the attempt to realize it not create an even worse disaster than the one in which we already find ourselves?

The first reply to these questions is that traditions that give individuals a home away from the 'home' offered by scientific–industrial societies and thus enable them to examine these societies rather than just living in

[42] Acupuncture can now be practised in California by people who are not MDs. But while an MD can practise outside his speciality, the acupuncturist needs an additional licence as a dietician if he wants to prescribe a diet, a further licence if he recommends teas, a still further licence if he massages. Money is of course the prime consideration. But there is also the idea that a person who has passed a pseudo-scientific education is in a better shape to judge things than a person who has not. The facts say otherwise.

them – such traditions already exist. Tribal traditions and traditions of non-Western empires survived the onslaught and the educational chauvinism of Western conquerors and have increased importance today when new classes and new races enter the daylight of civic life. They have not survived in their original form and some of them have to be reconstructed from meagre remnants – but there is enough material *and readiness* to build alternatives to the 'mainstream' of Western (which includes Communist Russian) culture. It is interesting to see how little use both liberals and Marxists have for these traditions. They examine them, they study them, they write about them, they 'interpret' them, they use them to bolster their own ideologies but they would never grant them a fundamental role in education, and they would never permit them to displace science from the central role it now assumes. This dogmatism is only rarely noticed, for nothing is now more popular than to praise primitive art, black music, Chinese philosophy, Indian stories and so on. What is not seen, not even by the concerned cultures and races themselves is that much of this so-called art was a science as well, it contained views of the world and rules for survival in it. What the interpreters show us today is a truncated version of such world views that makes them marvellous playthings for intellectuals (Marxists, psychoanalysts, etc.) but the very same intellectuals would reject the views the moment they assert themselves in full force: *'Racial equality' does not mean equality of traditions and achievements; it means equality of access to positions in the white man's society.*[43] It assumes the superiority of such societies and magnanimously grants permission to enter them *on their own terms*. A black man, an Indian can become a medical specialist, he can become a physicist, a politician, he can advance to positions of eminence and power in all these fields, but he cannot practise the 'scientific' subjects that are part of his own tradition, not even for himself and his co-traditionists. Hopi medicine is forbidden, for the Hopi as well as for anyone else. This attitude is shared by Marxists and by Liberals. It rests on an unexamined belief in the excellence of Western science and Rationalism. It rests on an unexamined belief in the excellence of the White Man's achievements in the domain of science and of knowledge in general.[44]

But these achievements – and with this I come to the second point – are

---

[43] I am now mainly describing the situation in the U.S.A. – but the ideology that underlies it is much more widespread.

[44] With women's liberation the situation is exactly the same. Most women fall over themselves to get access to male-defined positions so that they may be able to repeat, and, considering their verve, perhaps even amplify, male idiocy.

much smaller than is advertised. The technological circus is often redundant and alternative procedures often superior. Taken together with the considerations of the last section this means that traditions that differ from science are not pockets of wilful disregard of 'reality', but are either different ways of dealing with the real, or else accounts of parts of reality inaccessible to science. Besides, there is no reason why grown-up people who have traditions of their own should pay attention to what others call 'reality' especially in view of the fact that the scientific approach to reality respects only efficiency and theoretical adequacy no matter what damage this does to the spirit of man, while older traditions try to preserve the integrity of man and nature. There is much we can learn from non-Western traditions both in efficiency and in humanity.[45] And there is much we can gain from letting these traditions freely dwell in our midst instead of mutilating them by rationalist or Marxist 'interpretations'.

It is also clear that *personal liberty* will be much enhanced by the possibility of making a choice between different forms of life. Man, after all, should be able to do more than *imitate* his surroundings. He should also be able to *look through* them, to recognize their faults as well as their advantages and so become a conscious member of his tradition rather than a dummy swept along by the stream of history.[46] The presence of traditions different from his own enables him to acquire such consciousness and with it a certain amount of *intellectual* freedom. But dreams, peripheral thoughts, vague discontent now cease to be subjective afflictions as well and become possible avenues to reality. The avenues may be expanded into the public domain, they may become powerful practices that enhance man's *material* and *emotional* liberation, but they may also remain the private affair of a few. At any rate – every man has now a chance to combine his own self-liberation with objective social change and, thereby, with the liberation of others. Let us construct societies in which this combination becomes part of normal life!

---

[45] This has been realized by the Chinese Communists who forced hospitals and medical schools to use traditional medicine side-by-side with Western medicine. Similar means will have to be used by the democratic governments of the West, for there is no hope that 'the inner dialectics' of Western medicine will lead to an equally enlightened attitude. There is too much at stake, both financially and as regards the 'reputation' of Western science. Governments, however, have the duty to provide the best surroundings for their citizens that can be achieved with human means. Cf. *AM*, p. 50.

[46] To achieve a (partial) separation of man and his social habitat and thus to enable man to see the limitations of this habitat is one of the main purposes of Brecht's method of *Verfremdung* (distantiation).

4. Why Bother?

CS ask why an anarchist should pay attention to the irrationality of his reviewers (n. 218) and so assume that I am an anarchist. I am not aware of having made such a confession anywhere in my book. I say that the book was written 'in the conviction that anarchism . . . is excellent medicine for epistemology and the philosophy of science' (*AM*, p. 18) but, of course, I reserve myself the right not to act on this conviction and I frequently make use of this right: my private life and my book are two different things. I thought EA an interesting view, I asked myself how far it could be developed, I found that 'rationality' does not offer any arguments against it (and this means now both the rationality of Methodism and the 'rationality' of CS) and that science has many anarchistic features. For me this means that reason plays a much smaller part in the affairs of men, even in the affairs of science, than is assumed by our intellectuals and that it leaves room for the development of a great variety of different forms of life. Moreover, the world in which we live can lend substance to different approaches, and so it is up to us to be either orderly, i.e. to remain the 'bearers' of a well-defined tradition, or dada, i.e. to make the leap into the nothingness outside all traditions. I myself prefer an orderly way of life, partly for reasons of health, partly because I get easily confused though I am well aware and I have made use of the creative aspects of chaos. All these possibilities are beyond CS. They wear their minds upon their sleeve. They cannot imagine that a writer might write an account of one kind of life, live another, belong to a group that is connected with still another and make propaganda for a style that is different from all three. For them a human being is like a statue made of sand and dried: touch it at one point, and it disintegrates entirely.[47]

Secondly, an anarchist is of course not obliged to disregard argument. Arguments are not abolished, they are just restricted in their use. 'Anything goes', after all, means that argument also goes.

Thirdly, it is most interesting to see how rationalists react towards a product like my book. Rationalists want arguments and nothing else. My book is addressed to many people and so contains many different devices. There are arguments to make rationalists feel at home, arias in various keys to please the more dramatic reader, fairytales to capture the Romantic, there is rhetoric for those who like a hard-hitting debate with no holds

[47] A similar mistake is committed by those who regard me as a Popperian, or a former Popperian.

barred, there are personal remarks for people who rightly feel that ideas are made by men and that one understands them the better the more one knows about the minds that create them. Now the strange thing is that hardly anyone of the self-professed rationalists who have read the book have either *recognized* the arguments or *replied* to them, and the replies that were made are pathetic, to say the least (cf. Section 1 above). Moreover, they have often complained about the arias and the rhetoric (which are less than a tenth of the book) as if a writer had an obligation to please only them and nobody else. I don't recognize any such obligation, and even if I did it would be useless to act on it, for rationalists only rarely conform to the standards they try to impose on others.

Gellner's review, mentioned by CS, is a case in point. It appeared in the *British Journal for the Philosophy of Science*, the party organ of critical rationalism and edited by J. W. N. Watkins, the stern janitor of the Popperian temple. On the first page of the review Gellner confesses that the history and the philosophy of science (which are the areas covered by the journal) are beyond him, which means, on any reasonable account of the matter ('reasonable' in J. W. N. Watkins-terms, not mine) that he is incompetent to review *AM*. He is incompetent for other reasons as well. He has never heard of a *reductio ad absurdum* and he cannot read plain English.[48] In addition he does an interesting barber-job on his quotations. He quotes me, but omits a 'not' here and a 'but' there, thus turning statements of mine into their opposite. Now is this not an interesting fact? Is it not an interesting fact that critical rationalists under fire do not fight themselves but send illiterates to the trenches? No doubt they do this in order to preserve the dwindling number of hard-core rationalists who have not yet made fools of themselves in print. All this I thought was a most interesting fact, and I thought also that it deserved to be known to a wide audience – so I wrote my reply.

My reasons for replying to CS are somewhat different. By the time I met CS I had read a sizeable number of reviews. I discovered that almost all reviewers approach the book after a certain pattern. (1) They assume that a writer when presenting a position reveals his innermost thoughts: all books are autobiographical; (2) they extend this assumption to the more abstract parts of the book, thus reading indirect arguments as direct arguments, and so on; (3) they assume that what a person says or does forms a psycho-conceptual unit, a 'system' that can be explained in simple terms, so that blowing up the system is the same as blowing up the

[48] Cf. my reply pp. 140ff. above.

book; and (4) they can hardly ever read, or remember what they have read. I thought it interesting to give an account of this pattern as well, for it is fairly widespread and may explain certain features of scientific change and other types of change.

In the case of CS item (3) gets support from the reviewers' ideology, which is vulgar Marxism. Vulgar Marxism provides simple and manageable stereotypes, a kind of poor-man's thinking kit. Having formed a vague impression of *AM*, CS choose a suitable stereotype. Disregarding subtleties of argument, irony, indirect speech, and other features of civilized discourse, they notice ideas that fit the stereotype.[49] So now they have both my number and independent evidence for it. Next comes the moral evaluation or, to use simpler terms, abuse. Some of it is amusing: CS call me a class-peripheral parasite because of my jokes, which means that life must be pretty dull at the centre of classes. Much of the abuse is standard vulgar-Marxist vintage as spoken by Marxists in grade-B movies on the heroic deeds of the FBI. There is no analysis, no subtlety, there is

[49] To give their 'analysis' an air of subtlety they embroider it with 'evidence' derived from asides and personal remarks of mine. Some of their comments on these bits and pieces of Feyerabendiana (which for them become indicators of deep significance – cf. n. 215) reveal their usual inability to recognize the presuppositions of a statement, or of a position. For example, I recommend separation of state and science but I also recommend that the state intervene when science gets out of hand. CS (n. 203) infer that I have a 'peculiarly contradictory view about the state'. But intervention is compatible with separation if it tries to introduce it, or to protect it, or to restore it when it has been violated. In other comments CS reject my criticism because it goes against their party line. Thus I make some scathing remarks on the actions of left-wing students during the student revolution of the late sixties. CS do not ask what the actions were, they simply say that I lack political understanding (n. 206). Does this mean that 'radical action' is good and a critic of 'radical action' an idiot no matter what the details of the action? Does plain stupidity cease to be plain stupidity when practised by card-carrying 'radicals'? I had an opportunity to watch student radicals from close by, in Berkeley, London and Berlin. I often talked to them and I was appalled by their tactical innocence, their Puritanism, and their anti-humanitarianism. Especially in Berkeley opponents were not regarded as people who need *information*, they were despised and ridiculed. Political masturbation was the order of the day until Ronald Reagan, who knew very well how to make use of this incompetence, swept the whole affair away like a bad dream. To be a reformer it is not sufficient to have good intentions, one must also know a few things, one must be able to adapt abstract theory to concrete events and, above all, one must respect people, opponents included. Then CS dig up old events to complete their picture of me: I translated Popper's *Open Society*, his 'chief contribution to the Cold War' (n. 216). Again, there is no analysis (for example, I might have translated the book to expose Popper to German Marxists who were fascinated by his epistemology and needed waking up – this was *not* my reason, but it might have been), just the collection of brute facts and their addition to the stereotype (I wonder what CS would have said had they known that I loved to play tin soldiers when a child, was a lieutenant in the German army when somewhat bigger, and see at least one Samurai movie every week?). Marxism must be in a sad state if inarticulate fantasies can be regarded as knowledge and if a paste-job of stereotypes, uninterpreted facts, misread passages, and uncomprehended arguments can pose as analysis.

the blurred image of an opponent, a vague memory of party slogans and bang! Off goes the cannon. The cannon is not always CSs' own, they occasionally have to borrow, and they borrow from Gellner, the arch-reactionary. Now Gellner calls me a parasite because he thinks I want to exploit scientists but without adequately compensating them. What I actually suggest is that scientists be used, properly paid, and properly praised, but that they be no longer permitted to shape society in their image. The shaping of society is to be done by its citizens, not by power-hungry intellectuals. Gellner quotes the passage, omits from it the part referring to rewards and praise and launches on his accusation of parasitism. CS, who need but a single word to go off at a tangent, gratefully accept 'parasite' and add it to their stereotype. And so illiterate vanguard of the proletariat joins hands with illiterate rearguard of the reaction to come up with: PKF the parasite. Now let us look a little more closely at this charge, no matter how hilarious the reasons on which it is based. I am a parasite – but I am not aware that CS earn their income in any more strenuous way than I do. They lecture at universities, just as I do; they spend their time writing papers, reviews, just as I do; from their acknow-ledgements I gather they have typists to type their manuscripts and fall-guys for discussions, which is more than I have, for I type my own manuscripts and never bother anyone with my ideas – so what is this charge of parasitism based on? Does a person who lives a parasite's life cease to be a parasite when he starts writing Marxist essays, or are the reviewers simply irked by the fact that I make more money and take myself less seriously than they do? I do not know. What I *do* know is that it is interesting to see how Marxists, rationalists, and other intellectuals nowadays defend their respective chickencoops and to realize that their methods of defence accord with what I say about 'rational change' in *AM*. This is the reason why I replied to CS.[50]

---

[50] The Editor informs me that CS have made changes and cuts in their proofs. Naturally I have not been able to take account of these.

Chapter 4
From Incompetent Professionalism to Professionalized Incompetence – the Rise of a New Breed of Intellectuals

1. A Problem

Against Method was my first book and the first work whose reviews I studied in some detail. In the course of this study I discovered two things. Most reviewers are 'young' people whose career started one or two (academic) generations after the Kuhn–Lakatos era; and their reviews (with some rare exceptions here and there) have certain features in common. I found the features interesting, surprising and not a little disquietening, and I decided to take a closer look. Preliminary reports of what I found are my replies to Agassi, Gellner, Curthoys and Suchting, and Hellman[1]. When I wrote these replies I thought that I was confronted with individual incompetence: the learned gentlemen (and the one learned lady who joined the dance) were not too bright and rather badly informed and so they quite naturally made fools of themselves. Since then I have realized that this is a rather superficial way of looking at things. For the mistakes I noticed and criticized do not merely occur in this or that review, they are fairly widespread. And their frequency is not merely an accident of history, a temporary loss of intellect, it shows a pattern. Speaking paradoxically we may say that incompetence, having been standardized, has now become an essential part of professional excellence. We have no longer incompetent professionals, we have professionalized incompetence.

In what follows I shall try to do two things. First I shall try to exhibit

---

[1] Hellman's review appeared in *Metaphilosophy* 1978. This text was first published in *Philosophy of the Social Sciences* 1978. The reviews to which this note refers appeared in the autumn issue 1977 of the same journal. Section 3 has been rewritten and there are large omissions in other sections.

part of the pattern I am talking about. The three reviews of my book in this journal provide excellent material for such an attempt. Secondly, I shall try to explain how this pattern came into existence. My procedure in the first part will be to state theses and to illustrate them with examples from the reviews (H: Hattiangadi; K: Kulka; T: Tibbets). Each illustration will be accompanied by critical observations.

2. The Evidence

First thesis: Rational discourse is only one way of presenting and examining an issue and by no means the best. Our new intellectuals are not aware of its limitations and of the nature of the things outside.

Thus T undertakes his critique 'with understandable trepidation' because he feels that I am no 'longer amenable to normal standards of rationality and reasonableness'. 'It is impossible' he writes 'to determine when he is to be taken seriously and when he is merely absurd so as to shock and confuse non-Dadaists'. This is indeed impossible – given nothing but the criteria of the average philosopher of science. Using these criteria and nothing else it is also impossible to recognize irony, metaphor, playful exaggeration. Yet writers who have studied these categories, who have examined them in the work of others and who use them in their own creations are not at all at a loss, they make their judgements with minimal error and with perfect ease. True, their 'criteria' do not occur in the standard works on philosophy of science. But they can be learned, applied, refined. It is simply not true that a writer who leaves the domain of rational discourse ceases to make sense and that a reader who follows him is left without a guide, though it must appear so to those who have only read Popper and Carnap and never even heard of Lessing, Mencken, Tucholsky.

While T notices that I am not always engaged in a rational debate and infers that I am not always making sense, K recognizes no such distinctions. For him all statements are like 'the cat is on the mat'. It is hard to believe – but he actually fills three pages with arguments to prove that my dedication is false and that Imre Lakatos was not an anarchist (Gellner, at least, was content with a little sermon). Who had ever the slightest doubt about *that*? When *AM* was ready for print Imre Lakatos and I discussed various possibilities for a dedication. I considered: 'To Imre Lakatos, friend and fellow *rationalist*' – an ironical allusion to Lakatos' often voiced suspicion that I was a rationalist at heart and would recoil in horror if everyone became an anarchist (he was right). Next I considered dedicating the book to three alluring ladies who had almost prevented its completion. Lakatos approved for he knew two of them. Then I suggested 'To Imre Lakatos, friend, and fellow *anarchist*'. Lakatos said he was 'flattered' provided the comma I had put after 'friend' was removed (it wasn't).[2] Little did we know when amusing ourselves in this light-hearted fashion that such an obvious and transparent joke would one fine day be carefully analysed and be found to be wanting in truth content. Yet this is precisely what K does in his review. He knows that most people are by now acquainted with Lakatos' views. He assumes that I am aware of the fact. He notices a gap between what I say and what most people believe and give me credit for noticing the gap as well. So far so good. But now he assumes that the only way of closing a gap between two things is by argument and so he says that I 'try to *justify* my *charge*' (viz. that Imre Lakatos is an anarchist). Now it is quite true that my book contains a criticism of Lakatos' method of research programmes. One of the results is that while Lakatos abhors irrationality and anarchism he can exclude them only by using measures which are irrational by his own standards. This does not yet make him an anarchist, or an irrationalist: it makes him a rationalist who by misadventure ends up in irrationality. Accordingly I never *charged* Lakatos with anarchism. I merely *teased* him by calling him an 'anarchist in disguise' and by welcoming him as an (unwitting and unwilling) ally in the fight against reason. I closed the gap perceived by K not by an argument, or a presumed argument, I closed it by *mockery* and used arguments to give the mockery substance and not to 'justify' factual statements about Lakatos' philosophy. Sorry, dear Tomas, if all that has confused you but you should have read my announcement, in the preface, that I was going to present a *letter* (not an 'academic' book as H says) to *Lakatos* (not to illiterate pedants) and that my style would be that of a letter. Which brings me to the second thesis.

Second thesis: Although our new intellectuals extol the virtue of a rational debate they only rarely conform to its rules. For example, they don't read what they criticize and their understanding of arguments is of the most primitive kind.

In *AM* Chapter 2 I write: 'My intention is not to replace one set of general

---

[2] The relevant letters can be found in the Lakatos files at the LSE.

rules by another such set: my intention is, rather, to convince the reader that *all methodologies even the most obvious ones, have their limits.* The best way to show this is to demonstrate the limits and even the irrationality of some rules which she or he is likely to regard as basic. In the case of induction (including induction by falsification) this means demonstrating how well the counter inductive procedure can be supported by argument': counter induction, proliferation etc. are not introduced as new methods to *replace* induction or falsification, but as means of showing the limits of induction, falsification, meaning invariance and so on. Yet K says that I have a methodology and that 'anything goes' is its 'central thesis', T makes me defend a methodological pluralism with science at the surface and myth closer to the depths of understanding and H says that I want to replace induction by counter induction.

In the same chapter I emphasize that I don't pretend to know what progress is but use the account given by my opponents. K reads this as saying that 'there is no such thing as progress' and finds it 'strange' that the notion should still occur in my arguments (has he never heard of a *reductio ad absurdum*?) I emphasize that I am going to use rationalist procedures, argument included to make trouble for rationalists and not because I love argument, H says that I have 'not entirely "kicked away the ladder"' of rationalism.

These examples whose number could be easily increased show that the reviewers not only cannot read but are unfamiliar with an elementary rule of argumentation which was old hat when Aristotle wrote his *Topics* viz. that an argument is not a confession but an instrument designed to embarrass opponents. All that is needed for that purpose are (1) premises that are accepted by the opponent, (2) trains of thought that lead from the premises to conclusions which (3) are in conflict with beliefs of the opponent. Neither (1), nor (2), nor (3) entails that the author of the argument also accepts the premises, or the pattern of argumentation. Yet this is what the reviewers constantly assume, despite the additional help I give them in the form of explicit disclaimers.[3]

Here is still another example to illustrate the situation. K observes that Kuhn, Toulmin, Lakatos and I treat *historical* facts like 'sacred cows' although all of us are very critical towards *facts* on other occasions. I cannot speak for Kuhn etc. (though Elkana under whose benevolent supervision the paper was written should have known better in the case of

---

[3] For example, none of my critics seems to have noticed that I introduce 'anarchism' as a *medicine*, not as a final philosophy and that I envisage periods in which rationalism is preferable (end of Chapter 1).

Lakatos) but as far as I am concerned the situation is crystal clear. I analyse concrete events in ancient, early modern and late modern science and produce statements that can be regarded as descriptions of fact. Nowhere in the book do I regard these descriptions as 'sacred cows' i.e. as unalterable and absolute. Of course I use them to undermine the assumption that great science obeys universal standards but such use is perfectly compatible with their being hypotheses, or 'fairytales', as I occasionally call them. Critical rationalists constantly use hypotheses ('accepted basic statements') to criticize other hypotheses. I adopt their procedure (see above, on the elementary rule of argumentation) to make difficulties for their pet philosophy.

Besides, I do not confront rules and methods with a history that is separated from them. I invite the defenders of the rules to introduce them into the historical process they are interested in and predict that they will distort it in a manner most displeasing to them (the steps of classical ballet are beautiful to behold, but who would believe that one can use them to climb mountains?).

I said that much of the criticism reveals an inability to read and to understand simple arguments. Perhaps the fault does not lie with the reviewers? Perhaps the passages which I mention and which they overlook are tiny islands in an ocean of statements that point in a different direction? Let us see!

Third thesis: Historical studies are treated in a summary fashion or are altogether neglected even when they constitute the core of an argument.

K presents my case as follows: 'Since we cannot know anything for certain we cannot know anything at all and therefore all ideas have the same epistemic value' which means that I argue from uncertainty to ignorance. On another occasion K says that Lakatos and I have 'succeeded in demonstrating that there are historical counter-examples to all normative standards that were proposed as a universal method of science' and he also accuses us, in a passage already described, of using historical facts as 'sacred cows': I argue from uncertainty, I argue from 'sacred cows'; I say we cannot know anything at all, I 'demonstrate' – it is clear that K does not know what to do with the case studies.

He (and Elkana whose somewhat weighty hand becomes more perceptible on this occasion – why did he not come forth and take the blows himself instead of hiding behind a student?) has not read them either. Continuing the above passage he writes: 'Or, to come back to methodology: since there can be no perfect methodology, all methods are useless and therefore "anything goes"'. (T argues in a similar fashion.) But in my case studies I not only try to show the *failure* of traditional methodologies, I also try to show what procedures *aided* the scientists *and should therefore be used. I criticize some procedures but I defend and recommend others.* Thus while explaining the complex moves Galileo makes in his attempt to defuse the tower argument I make it clear that and why it was *reasonable* to proceed in this way and why the procedures recommended by some rationalists would have been disastrous. Towards the end of Chapter 17 I point out that it is such a study of concrete cases rather than the arid exercises of rationalists that should guide a scientist and I argue for an anthropological and against a logical study of standards. In Chapters 2, 12, 17 I show how and why the structure of knowledge and the laws of human development taken together with some simple cosmological assumptions favour a historical anthropological approach. None of these suggestions seem to have been noticed by my reviewers and this despite the fact that their discussion fills more than half of my book. All they notice are my somewhat ironical summaries and the only positive statement they find and immediately elevate into a 'central thesis', or a 'principle' of 'PKF's methodology' is the slogan 'anything goes'. *But 'anything goes' does not express any conviction of mine, it is jocular summary of the predicament of the rationalist*: if you want universal standards, I say, if you cannot live without principles that hold independently of situation, shape of world, exigencies of research, temperamental pecularities, then I can give you such a principle. It will be empty, useless, and pretty ridiculous – but it will be a 'principle'. It will be the 'principle' 'anything goes'.

Two sets of comments before proceeding to the next thesis.

(1) H has an interesting way of avoiding discussion of Chapter 17 which deals with incommensurability. He says that I agree with Giedymin that incommensurability is 'unclear and insufficiently precise', that I have many difficulties with it as I 'now admit', that I introduce incommensurability 'almost as an afterthought' and he therefore decides not to discuss the matter although it plays an important role in his account of 'where Feyerabend stands'. But when agreeing with Giedymin that incommensurability is 'unclear and insufficiently precise' I do not accept the criticism that goes with the phrase but argue that and why greater clarity and greater precision would be disastrous and the 'now' only shows that H has not read Chapter 17: for here I claim to have solved the difficulties I once (H's 'now') perceived. And as regards the 'afterthought' I

can well imagine why Chapter 17 appears to H in that light: it is mainly historical, the philosophical considerations are inseparably tied up with the case I am discussing, and history, anthropology and related matters are of course but 'afterthoughts' to our new breed of intellectuals.

H also says that *AM* fails because I regard theories as explanations and he somewhat sheepishly refers to work of his own where this assumption is dropped. But explanations play no role whatsoever in *AM*. They played a role once, in the paper H analyses in some detail, they no longer occur in *AM*: not every author carries the few things he finds with him as if they were his life savings. H points out that method cannot always be avoided – a singer, for example, must sing in a certain way, or he will get hoarse. Quite true! But such a 'certain way' cannot be caught in fixed and stable rules which is why there are various *schools* of singing and why a singer who violates basic rules shared by almost all schools may still sing better than his rule-bound colleagues (example: the late Helge Roswaenge). Finally, H tries to give his analysis depth by constructing an ideological background for my ideas. He thinks I am a Romantic.[4] I am indeed a Romantic but not in his sense. For him Romanticism is a yearning for old traditions and a love for imagination and emotion. I say that old traditions should be retained not because they are *old*, but because they are *different* from the status quo, because they permit us to see it in perspective and because many people are still interested in them and want to live accordingly. I also favour imagination and emotion but I don't want them to *replace* reason, I want them to *limit it*, and to *supplement it*. This, incidentally, was the intention of true Romantics such as Novalis and post-Romantics such as Heine. They are very different from the textbook Romantics H seems to have in mind (and who are mostly inventions of confused professors of literature).

(2) K, in perfect accordance with his inability to distinguish different types of statements (see above, material to thesis one) regards my joke 'anything goes' as a basic 'principle' of my 'methodology' – but he does not do a very good job with this 'principle' either. He says it entails 'don't select'. But if *anything* goes, then selecting also goes. He says the principle excludes scientific reason. But if anything goes, then scientific reason also goes. He says that people presented with the principle would refrain from reasoning. Those who cannot think unless guided by some leader, even a disembodied leader such as a methodological rule, would certainly be at a loss. But people of independent minds would use them

---

[4] Toulmin and others have tried to exhibit their wide reading by similar 'analyses'.

with renewed vigour. K says the Middle Ages were pretty intolerant. I didn't say they were not, I said that science has now taken over the mediaeval tradition of intolerance but rejected some of the marvellous ideas mediaeval philosophers were intolerant about and I suggest that these ideas be used. (T also remarks that witchcraft is as dogmatic as science; I wonder whether he has studied it. But I of course never denied that dogmatism can be found outside the sciences.) K says he is unaware of any impressive discoveries before the arrival of modern science and implies I expect things to be better in the absence of science. I quite believe that he has no idea as to what went on in the Stone Age, or in the 12th century – but since when has ignorance become an argument? Besides, I never said that narrowmindedness was the *exclusive* property of science. I only mentioned science as a *good example* that is *still with us*. K asks if I would take an understanding, sympathetic, attitude if all my papers were rejected. I could not care less for my principle is not (as his seems to be): I am printed, therefore I am. (Why, then, did I publish *AM*? To tease Lakatos, as I say in the preface.) K asks why I use aeroplanes and not brooms to get from one place to another. I have replied to questions of this kind in Appendix 4 and I repeat: because I know how to use planes but don't know how to use brooms, and because I can't be bothered to learn. K points out that I have not refuted all methodologies, that methodology is progressing (are his comments an example of that progress?) and that my theses may soon be overthrown. Well, I am waiting. . . .

Fourth thesis: Confronted with a challenge to basic beliefs (such as the belief that science excels above all other ways of understanding and mastering the world) our new intellectuals usually recite standard phrases from the rationalist breviary without argument. The more fundamental the challenge, the more sonorous the recitation.

One of the most frequent objections I meet is that there are 'contradictions' or 'inconsistencies' in my book. Almost everybody makes the remark, nobody explains why one should take it seriously. The accusation 'inconsistency!' – the mere *sound* of the word is supposed to act like a spell that paralyses the opponent. This is the first and most common example of a criticism by standard phrases.

Now to start with the number of contradictions in my book is much smaller than is believed by the critics and my explanations under thesis one and thesis two above have indicated why: the critics ascribe to me assumptions which I do not *accept* but *use* as part of my polemic against the rationalists. But, secondly, what is wrong with inconsistencies? True, there exist some rather simpleminded logical systems where a contradiction entails every statement but there exist also other systems, for example certain parts of science that do not have that property and then there are systems of logic, such as Hegel's where inconsistencies function as principles of conceptual development.[5] None of this seems to be known to the critics. The accusation of inconsistency therefore does not proceed from well thought out arguments. It is a knee-jerk reaction without intellectual content.

T assumes that I argue from fallibilism to the methodological inferiority of science. No such argument is found in my book. Nor have I ever said that science is inferior, methodologically, to other forms of knowledge. But I have opposed the wholesale condemnation of other forms on the grounds that they are not 'scientific' and I have criticized the image of science proposed by logicians and epistemologists. (This image is inferior both to science and to its alternatives.) Against my attempt to criticize judgements such as 'science is better than myth' T produces a series of slogans but no arguments. And as the slogans are found not only with him but also in other authors it is perhaps not uninteresting to examine them.

What gives science its privileged status, says T, is the self-critical, self-correcting character of scientific inquiry. Such a statement makes nice reading for rationalists – but is it true, is it desirable if true, does it distinguish science from other enterprises?

To take the last point first: science is self-corrective, witchcraft is not. We shall soon see that T has not examined the matter too carefully in the case of science – but where did he get his information concerning witchcraft? What particular witchcraft doctrine has he examined to have come to the conclusion that witchcraft is not self-corrective? Science is now accessible to everyone, one can study it in the open air as it were though the study is difficult and correct results are not easily obtained. But there is now hardly any reputable school of witchcraft left at the places T

---

[5] Popper's criticism of Hegel proceeds from the assumption that Hegel's logic contains the simpleminded logic in which a contradiction entails every statement while Hegel from the very beginning emphasizes the difference between his logic and the 'formal logic' of his contemporaries. Hence, all Popper can say is that Hegel's logic is not the logic he loves – he cannot say that it is inadequate. Carnap's criticism of Heidegger is simpler but of exactly the same kind. He shows that the negation sign of propositional logic is not used as 'Das Nichts' in Heidegger's 'Das Nichts nichtet'. That would be an objection if Heidegger had intended to describe the life of the negation sign – which he certainly did not.

frequents while the anthropological studies from which he might have obtained his information cover too short a period to be decisive. Besides, they show witchcraft in a state of dissolution. Also anthropologists are notoriously bad methodologists and so their results must be checked very carefully just as one must check what historians of *science* offer as the main features of scientific development. Which anthropologists did T read, how did he do his checking, what particular form of witchcraft does he base his assertion on? We receive no answer. And from what he says on other occasions we must infer that he just repeats rationalist *gossip* without having examined its validity. But then we obviously don't have an argument, we have another knee-jerk reaction, another pious phrase.

That the pious phrase is incorrect in many cases T seems to have in mind is shown by a study of the development of Church Doctrine. This enterprise seeks out error, defines it, eliminates it and so constantly improves basic theory. The views on angels held by St. Thomas are different from the views of angels held by St. Augustine, they are the result of a discussion that takes St. Augustine's views into account, they are the result of a self-correcting debate. T may of course object, and many rationalists have objected that theology has not *improvement*, but only *change*. I do not want to argue the point, I rather ask what gives them the assurance that the same is not true of science? Once scientists believed in the ether, then they eliminated it. A non-ether period followed an ether period. If we want to say that this sequence is accompanied by the elimination of error then we must be able to say that the situation is now better than it was before. And if we want to postulate error-elimination for all science, then we must have universal standards to make such judgements possible. But all the universal standards that have been proposed so far are in conflict with scientific practice (they are not just *fallible*, they are often inapplicable, or invalid). With this a main rationale for the belief in the self-correcting character of science disappears.[6] What other rationale has T to offer?

Thirdly, is self-correction of the kind envisaged by T *desirable*? Aristotle thought it was not. He gave not only arguments to that effect (cf. his arguments against Parmenides), he also constructed a cosmology, a physics, an astronomy, a psychology, a theory of politics, an ethics, a theory of drama that conformed to his idea that while science may study

---

[6] And as regards the *fact* of self-correction – need I remind T that many parts of science have now become businesses when the aim is no longer to find truth by self-correction (if that ever was the aim) but to keep the money coming. Disasters are presented as successes to get closer and closer to the aim, lying and self-deception are the order of the day.

and correct *local errors* it must leave the *general outlines* of the world unchanged. These outlines are determined by the nature of man and by his place in the universe. Knowledge depends on them; it does not depend on the dreams of small groups of intellectuals. Aristotle also developed a theory of change and observation that fits the facts, explains why such a fit gives knowledge and aids in the removal of error. What argument does T have against such a procedure? It is not 'scientific' in his sense, agreed, but is it worse? True, Aristotle was rejected in the 16th and 17th centuries – but why should we repeat the judgement of 17th century chauvinists? They could quote some reasons in astronomy and physics (which were by no means as unambiguous as one wants to make them today – cf. Part One) but there were no reasons in psychology, physiology, medicine. Here Aristotle was still used with success by outstanding researchers such as Harvey.[7] In the biological sciences Aristotle's general laws of motion were applied until late into the 19th century – and with excellent results. And then there is drama which according to Aristotle provides an *explanation* of apparently accidental historical events (it is sociology and as such superior to history), there is his politics, the history of ideas and so on: Aristotle is still with us and has to be. But all T seems to know in a very superficial way are certain episodes from astronomy (which for him are peripheral to science anyway – see below) and that settles the matter. Besides, not only Aristotle failed – modern science constantly fails and is still retained. All these problems are disregarded in the bland use of the phrase the 'self-correcting nature of science'.

Science, T says, is not only self-correcting, it is also successful in terms of prediction and 'credibility over time'. Credibility one may grant, for science is indeed most credible to the faithful. It is different with the *reasons* for this credibility. The main reason, says T, is 'successful prediction'. 'I cannot agree' he intones 'that it is my subjective and arbitrary preference or social conditioning that makes me take my sick child to the pediatrician rather than to the witch doctor. I go to the former because I believe – on the basis of individual and community experience – that the judgements of the former possess a greater likelihood of being true in both the short and long run than those of the latter'. And, 'given a set of symptoms such as a sore throat, high temperature and red spots then measles vaccine is more likely to be effective than the laying of hands and incantations'. He admits that science may occasionally fail –

---

[7] Cf. Walter Pagel, *William Harvey's Biological Ideas*, New York 1967.

it is not infallible – but he asserts that it is *more likely* to succeed. This is what he says. And considering what fuss he makes about the need to base scientific judgements on controlled experiments one would assume that he has carried out such experiments, or read of experiments where the successes of expert witch doctors (*not* of incompetents) are compared with the successes of expert pediatricians (*not* of incompetents) or with the effect of vaccines. Why did he not describe these experiments? Has he tried a witch doctor? ('Individual' experience) – and what is the 'community' he refers to ('community' experience)? Why such a chaste reticence in regard to the sources of his conviction? Or should it be the case that there are no experiments? That the 'individual and community experience' he refers to is an experience of faith and of slogans rather than of facts and that T's attitude is therefore the result of social conditioning, just as I have said of other intellectuals?

(I, on the other hand, have detailed 'community experience' and 'individual experience' with faith healers, acupuncturists, witches and other disreputable people. I had many opportunities to compare their efficiency with that of 'scientific' doctors and I have avoided the latter like the plague ever since.)

T's brief remarks about astrology exhibit the same self-assured ignorance (and here he is in good company – for the encyclical against astrology that appeared in the *Humanist* and was signed by 186 'scientist', 18 Nobel prize winners among them is a paradigm of proud illiteracy). Astrology, T says, is accompanied by absolute certainty. Now we have already seen that that alone does not yet make it suspect (see above, remarks on Aristotle). But the assertion is not true either. Astrology can be and has been revised to take new discoveries into account (example: revisions after the discovery of Neptune and of Pluto). One of the major revisions in the past was carried out by Kepler who defended sidereal astrology, opposed tropical astrology (does T know what these terms mean? Has he ever heard of them?) and assembled evidence for statistical predictions by the former.

T is not only ignorant of the non-scientific areas he condemns, he has also some very strange ideas about science. He says I generalize from the tower experiment and assume that all experiments are of the same logical type. There is not a trace of such a generalization in my book. What I do say is that arguments (remember that I call the tower-case an *argument* and *not* an *experiment*) involving a change of concepts have some common features that can best be found by an analysis of the tower argument. I would add that the 'experiments' T has in mind assume that one knows what concepts to use and therefore presuppose the type of argument I examine. Now for T science is statistical methods and controlled experiments and he complains that I 'seldom if ever descend' to that 'level'. Indeed, I don't descend to that level because I am more interested in the arguments and experiments that bring about fundamental change. Michelson's experiment, the experiment of Reines, of Weber, most microphysical experiments are of this type and so is all astronomy. I can easily see how a discussion of such events will seem 'to be totally divorced from the actual, ongoing context of inquiry' for somebody who objects to astronomy as a paradigm of science and whose concept of science fits best the slums of statistical studies in sociology. But let us not quarrel about preferences! I am content to have shown, partly by my own research, partly with the help of others that Galileo, Einstein, Kepler, Bohr etc. did not proceed in accordance with universal standards and I can well do without support from the types of research T is interested in.

A few minor points: T says that there is no sense in saying that evidence is contaminated if one does not grant the existence of healthy evidence. I agree and I have therefore suggested different terminology.[8] This T does not seem to have noticed. T says Don Juan has no understanding of the chemical composition of drugs and asserts that he therefore misses something. But that is the point at issue: do we know anything about drugs when we know their chemical composition? T regards my book less as a contribution to the philosophy of science than as a symptom which is fine with me considering what the philosophy of science seems to have become: a collection of pedantic comments based on rather simpleminded ideas about human discourse and surrounded by phrases which sound good in the ears of the faithful but for which no arguments are given and which reveal a gigantic ignorance of the things judged. Once philosophers of science were intelligent and well informed people and philosophy of science a rather interesting subject. What has happened since then? How can the deterioration be explained? Let us see!

3. Why some Modern Philosophers of Science are so much more illiterate than their Predecessors: Observations on Ernst Mach, his Followers and his Critics

Modern philosophy of science arose from the Vienna Circle and its attempt to reconstruct the rational components of science. It is interesting

---

[8] Cf. also what I say about theory-ladenness in n. 9 of my reply to Curthoys and Suchting.

to compare its approach with that of earlier philosophers, for example Ernst Mach.

Ernst Mach was a scientist. He was an expert in physics, psychology, physiology, the history of science and the general history of ideas. Ernst Mach was also an educated man. He was familiar with the arts and the literature of his time and he was interested in politics. Even when already paralysed he had himself wheeled into a session of parliament to cast his vote in connection with workers' legislation.

Ernst Mach was not satisfied with the science of his time. As he saw it science had become partially petrified. It used entities such as space and time and objective existence but without examining them. Moreover philosophers had tried to show and scientists had started believing that these entities could not be examined by science because they were 'presupposed' by it. This Mach was not prepared to accept. For him every part of science, 'presuppositions' included was a possible topic of research and subjected to correction.

On the other hand it was clear that the correction could not always be carried out with the help of the customary procedures, which contained some ideas in a way that protected them from difficulties. It was therefore necessary to introduce a new type of research based on a new cosmology. Mach gave a rough outline of what it would assume and how it would proceed.

According to Mach science deals with *elements and their relations*. The nature of the elements is not given but must be discovered. Known things such as sensations, physical objects, systems of physical objects in space are combinations of elements. The combinations may reproduce the old distinctions, but they may also lead to arrangements of an entirely different kind; for example, they may lead to an interpenetration of 'subject' and 'object' in the old sense. Mach was convinced that the old distinctions were inadequate and had to be given up.

Mach's conception of science has two features that distinguish it from what philosophers of science think about the matter today.

First, Mach was critical towards science *as a whole*.[9] Modern philosophers occasionally make a big show of their independence and their expert knowledge by criticizing particular scientific theories and suggesting minor changes. But they would never dare to criticize science as a whole. They are its most obedient servants. Secondly, Mach criticized

[9] Cf. his debate with Planck, reprinted in S. Toulmin (ed.) *Physics and Reality* New York 1965.

scientific ideas not by comparing them with external standards (criteria of meaning or demarcation) but by showing how *scientific research itself* suggested a change. For example, methodological principles were examined not by consulting an abstract and independent theory of rationality but by showing how they aided or hindered scientists in the solution of concrete problems. (Later on Einstein and Niels Bohr developed this procedure into a fine art.)

A third interesting feature of Mach's 'philosophy'[10] was its disregard for distinctions between areas of research. Any method, any type of knowledge could enter the discussion of a particular problem. In building up his new science Mach appealed to mythology, physiology, psychology, history of ideas, history of science as well as to the physical sciences. The *magic world view* which he received from Tylor and Frazer dissolves the distinction between subject and object without ending up in chaos. Mach did not accept this world view, but he used it to show that the 19th century idea of objective existence was not a *necessary* ingredient of thought and perception. His detailed studies in the physiology of the senses showed him that it was not *adequate* either. Sensations are complex entities containing 'objective' ingredients, 'objects' are constituted by processes (Mach bands, for example) which belong to the 'subject', the boundary between subject and object changes from one case to the next: for us it lies at our fingertips, for the blind man using a stick it lies at the end of the stick. The *history of science* and *physics* showed that 'objective' theories such as Newton's theory of space and time and atomism were in trouble precisely because of their objectivist features. On the other hand, there existed theories of a different kind such as the phenomenological theory of heat which were successful though not based on material substances. Having started with a classification of sensations (cf. Mach's *Theory of Heat*) such theories suggested to Mach that *at least at this stage of research* elements could be identified with sensations. For the time being Mach's new science could therefore be developed from two assumptions viz.

(1) the world consists of elements and their relations. The nature of the elements and of the relations as well as the way in which things are built up from them is to be determined by research using the concepts that seem most economical at a certain stage of science and

(2) elements are sensations.

[10] I put the word in quotation marks because Mach always refused to be regarded as the proponent of a new 'philosophy' – which agrees with the above account of this research practice.

This is how Mach combined the information provided by different fields to give shape to his own idea of research.[11]

Mach's idea of research was more comprehensive than that of his contemporaries and certainly of all his philosophical successors. Before it was taken for granted that not all parts of science could be examined by scientific means. Space, time, observer independence were thought to be beyond the reach of (scientific) argument. Now there were means of criticizing not only these ideas but the very standards of research: no standards can guide research without being subjected to the control of research.

It is interesting to see how later 'scientific' philosophers changed this rich and fruitful point of view. Mach's attempt to make research more comprehensive so that it could deal with 'scientific' as well as 'philosophical' questions remained unnoticed both by his followers and by his opponents. What they did notice were his assumptions and hypotheses and those they turned into 'principles' of precisely the kind Mach had rejected. The theory of elements became a 'presupposition', the identification of elements and sensations a definition and relations between concepts were imposed in accordance with some rather simpleminded rules, they were no longer determined by research. The construction of conceptual systems having such rules and such principles as their boundary conditions now became *the* task of a new and rather aggressive discipline – the philosophy of science. With this the old dichotomy between philosophical speculation and scientific research which Mach had tried to absorb into science reappeared – but it was a very impoverished and illiterate philosophy that took the place of its glorious ancestors. Being contemptuous of earlier ideas the new philosophers lacked perspective and soon repeated all traditional mistakes.[12] There arose then again two ways of dealing with general problems such as problems of space, time, reality and related problems viz. *the way of the scientists* and *the way of the philosophers*.

---

[11] Note that the second assumption above is a *hypothesis* and not a 'presupposition' of research. It is comparable to the assumption of 'objectivist' scientists that the ultimate building blocks of matter are elastic spheres, like billiard balls. It gets research started, it is not an unchanging standard of its adequacy. Having criticized the standards and 'presuppositions' of the science of his time Mach was not going to replace them by some other dogmatism (this becomes very clear from his notebooks).

[12] A predicament they shared with the enlightenment – except that the writers of that period *invented* their philosophy while the members of the Vienna Circle just *copied* the distorted ideas of their great predecessors. Also philosophers of the enlightenment dealt with

A *scientist* starts with a bulk of material consisting of diverse and conflicting ingredients. There are theories formulated in accordance with the highest standards of rigour and precision side by side with unfounded and sloppy approximations,[13] there are 'solid' facts, local laws based on some of these facts, there are heuristic principles, tentative formulations of new points of view which partly agree, partly conflict with the accepted facts, there are vague philosophical ideas, standards of rationality and procedures that conflict with these. Being unable to make such material conform to simple views of order and consistency the scientist usually develops a *practical logic* that permits him to get results amidst chaos and incoherence. Most of the rules and standards of this practical logic are conceived *ad hoc*, they serve to remove a particular difficulty and it is not possible to turn them into an organon of research. 'The external conditions' writes Einstein[14] 'which are set for [the scientists] ... do not permit him to let himself to be too much restricted, in the construction of his conceptual world, by the adherence to an epistemological system. He therefore must appear to the systematic epistemologist as a type of unscrupulous opportunist . . .' And Niels Bohr 'would never try to outline any finished picture, but would patiently go through all the phases of a problem, starting from some apparent paradox, and gradually leading to its elucidation. In fact he would never regard achieved results in any other light than as starting points for further exploration. In speculating about the prospects of some line of investigation, he would dismiss the usual considerations of simplicity, elegance, or even consistency with the remark that such qualities can only be properly judged after the event . . .'[15] It is of course possible to describe particular cases but the only lesson we can draw from the descriptions is cautionary: never expect a

---

ethics, aesthetics, theology, they founded a new anthropology and considerably widened the horizon of their contemporaries. Nothing even slightly comparable is offered by the new 'scientific' philosophy that came out of the Vienna Circle (and Popperianism) and which is mostly concerned with the physical sciences and some distorted images of man, as we have seen. Any extension beyond these boundaries is second-hand imitation of earlier views and shares the superficiality of such imitations. Characteristic for the enterprise is a schoolmasterly tone which occurs wherever imagination has ceased to work and has been replaced by routine responses. A superficial comparison between Popper and, say, Lessing shows the difference between true enlightenment and the slavish imitation of its outer form. (Kant who wanted to become famous and who knew that schoolmasters are more readily accepted than independent minds changes his style in midlife. And he was right: the three Critiques became a great success.)

[13] Cf. the account of ad hoc approximations in *AM*, p. 63.

[14] *Albert Einstein: Philosopher-Scientist*, ed. Schilpp, New York 1951, pp. 683ff.

[15] L. Rosenfeld in *Niels Bohr, his Life and Work as seen by his Friends and Colleagues*, ed. S. Rosenthal, New York 1967, p. 117.

clever trick, or a 'principle', that helped on one occasion to be useful on another. One outstanding feature of scientific research especially of the kind envisaged by Mach is its disregard for established boundaries. Galileo argued as if the distinction between astronomy and physics which was a basic presupposition of the knowledge of his time did not exist, Boltzmann used considerations from mechanics, the phenomenological theory of heat and optics to determine the scope of the kinetic theory, Einstein combined specific approximations with a global and very 'transcendental' survey of physical world views, Heisenberg got some of his basic ideas from the *Timaios* and, later on, from Anaximander. Metaphysical principles are used to advance research, logical laws and methodological standards are suspended without much ado as constituting undue restrictions, adventurous and 'irrational' conceptions abound. The successful researcher frequently is a literate man, he knows many tricks, ideas, ways of speaking, he is familiar with details of history and abstractions of cosmology, he can combine fragments of widely differing points of view and quickly switch from one framework to another. He is not tied to any particular language for he may speak the language of fact and the language of fairytale side by side and mix them in the most unexpected ways. And, mind you, this applies both to the 'context of discovery' *and* to the 'context of justification' for examining ideas is just as complex an activity as introducing them.

The strife about the kinetic theory of matter towards the end of the last century and the rise of the quantum theory are excellent examples of the features I have just described. In the case of the quantum theory we have: classical celestial mechanics, classical electrodynamics and the classical theory of heat. Sommerfeld and Epstein strained the first and the second to the limit by supplementing them with a 'fourth Keplerian law' viz. the quantum conditions. Their successes suggested that quantum mechanics might be developed out of classical theory without too much change. On the other hand the original considerations of Planck, generalized by Poincaré, seemed to indicate that fundamental ideas such as the idea of a trajectory were inherently problematic. Einstein, recognizing their problematic character worked almost entirely with approximations and inferences therefrom and his results (photoelectric effect; statistical studies) had only a limited application: they could not explain the laws of interference. They even seemed to clash with experiments and received scant attention until Millikan showed the correctness of some of the predictions made. Working with approximations then became *the* method of the Copenhagen school. The method was disliked and not well understood by physicists of Sommerfeld's persuasion but explained the limited applicability of even the most subtle mathematical instruments. And as a great and troublesome river disposes many strange objects on its shores, in the very same fashion the great and troublesome river of pre-1930 quantum mechanics produced numerous precise but little understood results, both in the forms of 'facts' and in the form of 'principles' (Ehrenfest's principle of adiabatic change being one of them).

The *way of the philosopher* is very different – there could not be a greater contrast. There are some general ideas and standards which are spelled out in detail and there are the principles of the logic chosen. There is hardly anything else – a consequence of the 'revolution in philosophy' initiated by the Vienna Circle. The logic used was of course discussed and it changed, for logic is a science like every other science – but only its most pedestrian parts entered the philosophical debate. Thus we have not only a separation between science and philosophy, but a further separation between a scientific ('mathematical') logic and a logic for philosophers. It is as if scientists used not the most advanced mathematics of their time but some backward idiom and tried to formulate their problems in its terms. Research of the philosophical type then consists in proposing ideas that fit the boundary conditions, i.e. the standards and the simple logic chosen.

Such ideas clearly are both too wide and too narrow. They are too wide, because the contemporary knowledge of facts is not taken into account (a purely philosophical theory of walking is bound to be too wide because it does not consider the restrictions imposed by physiology and landscape). And they are too narrow because the restricting standards and rules are also unaffected by that knowledge (a purely philosophical theory of walking is too narrow because it imposes restrictions not parallelled by the vast possibilities of human motion). It is this last feature that makes a philosophical criticism so dreary and repetitive. While a good scientist objects to 'making a good joke twice'[16] a philosopher insists on standard arguments against standard violations of standard standards. Exclamations such as 'inconsistent!'; 'ad hoc!'; 'irrational!'; 'degenerating!'; 'cognitively meaningless!' recur with tiring regularity. Illiteracy, however, not only does not matter, it is a sign of professional excellence. *It is required*, not just tolerated. All the distinctions of the discipline (context of discovery/context of justification; logical/psychological;

---

[16] Einstein's reply to the question why he did not stick to the philosophical ideas that led him to special relativity.

internal/external; and so on) have but one aim: to turn incompetence (ignorance of relevant material and lack of imagination) into expertise (happy assurance that the things not known and unimaginable are not relevant and that it would be professionally incompetent to use them).

The much admired addition of modern formal logic to philosophy has encouraged the illiteracy by providing it with an organon. More than anything else it enabled the barren fathers of positivism to deny their shortcomings and to assert, not without considerable pride, that they were not concerned with the advancement of knowledge but with its 'clarification' or its 'rationality'. Even critics did not try to re-establish contact with the practice of science,[17] they merely tried to free the suggested 'reconstructions' from internal difficulties.[18] The distance between scientific practice and the philosophy of science remained as large as ever. But this deficiency, this astounding unreality of the enterprise had already been turned into an asset: differences between reconstructions and actual science were regarded as faults of *science*, not as faults of the reconstructions. Of course, nobody was bold enough to play this game with physics (though there were some who derived much mileage from conflicts *inside* physics such as the conflict between Bohr and Einstein) – but if the trouble came from a less adored science then the verdict was clear: off with her head! While Mach's criticism was part of a *reform* of science that *combined criticism with new results* the criticism of the positivists and of their anxious foes, the critical rationalists proceeded from some frozen ingredients of the Machian philosophy (or modifications thereof) that could no longer be reached by the process of research. Mach's criticism was dialectical and fruitful, the criticism of the philosophers was dogmatic and without fruit. It mutilated science instead of making it grow. This started the trend whose late children we have now before us.

It is interesting to compare the two procedures in a concrete case.[19]

Mach's idea of a science all of whose standards and principles are under its own control was realized, in different ways, by Einstein and by Bohr. Interestingly enough both scientists (and some of their followers such as Max Born) regarded themselves as dilettantes, they defined and

[17] Lakatos, of course, tried to find a connection, but he came later and succeeded in making only *verbal* contact; cf. *AM*, pp. 196ff.

[18] Thus Popper's theory of falsification concerns an improvement of *confirmation logic*, not of science. The same is true of his theory of verisimilitude.

[19] In the Vienna Circle only Neurath had a clear conception of the properties of scientific research (as opposed to philosophical analysis). The difference between the two modes is well explained in Ayer's criticism of Neurath in his *Foundations of Empirical Knowledge*.

approached their problems independently of existing standards. They had no compunction about mixing science and philosophy and so about advancing the cause of both. Einstein's philosophical bent becomes clear from the way in which he arranges his material, Bohr's philosophy is an essential element of the older quantum theory.[20] It is true that Mach was severely critical of some later consequences of Einstein's research – but one should examine his reasons before concluding that Einstein was outside Mach's research programme. Nobody has so far paid attention to Mach's remark, contained in his criticism, that his investigations in the physiology of the senses had led him to results different from those ascribed to relativity. This establishes a connection with Mach's earlier analysis of space and time and indicates that he did not object to the new *theory* but to its *reification* by Planck and von Laue. For here relativity was used to support the very same naive and inarticulate notion of reality Mach had objected to and had started to examine. The examination was continued by the quantum theory which gave new content to the notion of an element, revealed new and complex relations between elements, and so modified our idea of reality. All this occurred in the twenties and thirties. What did the philosophers have to offer, during this time and after?

They had very little to offer in the case of relativity. They watched the development from the wings, applauded it, and 'clarified' it, i.e. described it as an instance of problem solving in their sense. The 'clarifications' created some interesting philosophical myths. For example, it created the myth that Einstein advanced by eliminating metaphysics, or by eliminating ad hoc hypotheses, or because he was an operationalist, or because he took refutations seriously. Zahar's remark that special relativity was no advance at all is the latest and most amusing myth of this kind.

The situation was different in the case of the quantum theory and its conception of 'reality'. While the quantum race was on the Vienna Circle changed from sense data languages to physicalistic languages. The change was as arbitrary as the choice of sense data had been in the first place. Sense data were removed because one returned to an interpretation of science the idea of sense data had been supposed to test. And one returned to such an interpretation not because a test was performed and failed, one was not even aware of the test function of sense data in Mach's philosophy, one simply remembered some principles of the science he wanted to improve and used them as arguments against carrying out the

---

[20] For details cf. my paper on Bohr in *Philosophy of Science* 1968/69.

improvement.[21] This rather unreflected turnabout occurred at precisely the time when the idea of objective existence was examined by physicists and replaced by a more complex account of reality. The event had no effect whatsoever on the debates between physicalists and the defenders of sense data and one can see why. It was regarded not as introducing different, more complex and more realistic arguments about the issue but simply as a technical but philosophically inferior version of an examination of the second, the philosophical kind. This becomes very clear from Popper's account of the matter. Writing more than twenty years later he complains: 'Without any debate over the philosophical issue, without producing any new argument, the instrumentalist view has become an accepted dogma'.[22] For him the detailed physical arguments, the many attempts to escape 'instrumentalism' as he calls the final position of the Copenhagen school simply do not exist. The tendency to 'translate' cosmological assumptions into the 'formal mode of speech' and so to conceal their factual content aided this blindness and the resulting rigidity of the philosophers' approach. Thus Popper, in the essay already quoted[23] removes 'essentialism' and introduces 'realism' by referring to the 'fact' (as he calls it) 'that the world of each of our theories may be explained by further worlds . . . described by further theories'. This, of course, is his model of science in which rejection of ad hoc hypotheses plays an essential role. The model breaks down in a finite world – but the breakdown will never become visible to a philosopher who hides factual assumptions behind 'logical' principles and 'methodological' standards. This is how complex problems needing unusual ideas and unusual minds were turned into trite puzzles that were then explained at great length and solved with a great show of intellectual effort.[24] And this is how the reality conception of classical physics could stage a comeback in philosophy after it had been defeated by scientific research.

The writers of the Vienna Circle and the early critical rationalists who distorted science and ruined philosophy in the manner just described belonged to a generation still vaguely familiar with physics. Besides, they started a new trend, they did not merely take it over from more inventive predecessors. They *invented* the errors they spread, they had to *fight* to

---

[21] The arbitrary nature of the change is clearly realized by Carnap who in his *Logical Syntax of Language* and earlier papers makes the choice of sense data vs. physicalistic languages a matter of *convenience* ('principle of tolerance').

[22] *Conjectures and Refutations*, London 1963, pp. 99f.

[23] *op. cit.*, p. 115.

[24] Many papers of positivists and critical rationalists can be summarized in a few lines.

get them accepted and so they had to possess a modicum of *intelligence*. They also suspected that science was more complex than the models they proposed and so they worked hard to make them plausible. They were pioneers, even if only pioneers of simplemindedness. The situation is very different with the new breed of philosophers of science that now populate our universities. They received their philosophy ready made, they did not invent it. Nor do they have much time or inclination to examine its foundations. Instead of bold thinkers who are prepared to defend implausible ideas against a majority of opponents we have now anxious conformists who try to conceal their fear (of failure, of unemployment) behind a stern defence of the status quo. This defence has entered its epicyclic stages: attention is directed to details and considerable work is done to cover up minor faults and deficiencies. But the basic illiteracy remains and it is reinforced for hardly any one of the new breed possesses the detailed knowledge of scientific procedure that occasionally made their ancestors a little hesitant in their pronouncements. For them 'science' is what Popper or Carnap or, more recently, Kuhn say it is – and that is that. It is to be admitted that some sciences, going through a period of stagnation now present their results in axiomatic form, or try to reduce them to correlation hypotheses. This does not remove the stagnation, but makes the sciences more similar to what philosophers of science think science is. Having no motivation to break through the circle and much reason (both emotional and financial) to stay in it philosophers of science can therefore be illiterates with good conscience. Small wonder that intelligent criticism is hard to find . . .

Postscript

H and T have replied to my remarks (*Phil. Soc. Sc.* 1978, 55ff). T wants me to give a more detailed account of the literature that made me criticise his idea of myth and of the experiences which made me doubt the competence of scientific medicine. But in his review he asserted the supremacy and self correcting character of science without the hint of any argument, i.e. on an even more slender basis than I used to criticise him. If he was satisfied with his statements *then*, the why is he not *now* satisfied with mine? T also asks: 'If *he* had a child diagnosed with Leukemia would he look to his witchdoctor friends or to the Sloan Kettering Institute?' I can assure him that I would look to my 'witchdoctor friends' to use his somewhat imprecise terminology *and so would many other people in*

California whose experience with scientific medicine has been anything but encouraging. The fact that scientific medicine is the only existing form of medicine in many places does not mean that it is the best and the fact that alternative forms of medicine succeed where scientific medicine has to resort to surgery show that it has serious lacunae: numerous women, reluctant to have their breasts amputated as their doctors advised them went to acupuncturists, faithhealers, herbalists and got cured. Parents of small children with allegedly incurable diseases, leukemia among them, did not give up, they consulted 'witchdoctors' and their children got cured. How do I know? Because I advised some of these men and women and I followed the fate of others. The drawbacks of modern scientific medicine are discussed with ample documentation in I. Illich *Medical Nemesis* and, from a different point of view, in Coulter, *Divided Legacy*. The latter work shows how theoretical ideas that were imported into medicine without regard to medical experience proper eliminated valuable practices and lowered the success rate of medicine as a whole. As regards the ability of 'witch doctors' to further explain any discovery they may have made embroiled as they are in a theory of evil spirits I refer T to the history of classical withcraft from the *Canon Episcopi* to the *Malleus Maleficarum* which is the history of a progressive (in the sense of Lakatos) development on the basis of a detailed theory of demons. St. Thomas Aquinas' theory of the implicit pact plays an important role in this development and considerably increases the content of the theory (*Summa Theol.* II/ii, section 96). For the shape of the theory in the 17th theory cf. the essay by Trevor-Roper, for the development e.g. the treatise by Hansen together with Hansen's source book and the material at the Dean White Collection of Cornell University Library. T's contention that the contribution of non-scientific forms of life to knowledge is to *stumble on* interesting casual relations in places where scientists would not have looked without the ability to *explore* them once found is refuted by the material in Levi-Strauss *The Savage Mind* esp. Chapter 1, by Stone Age Astronomy (cf. the work of Thom, Hawkins, the discussions at the conference about the place of astronomy in the ancient world, edited by the British Academy) taken in conjunction with the interpretation of A. Marshack (*Roots of Civilization*) and de Santillana – von Dechend (*Hamlet's Mill* with comprehensive bibliography), by the history of acupuncture which indeed started as a chance discovery but was developed from there into a whole medical system (cf. the work by Needham, especially the fortcoming vls. on biology), by what is now known about Polynesian astronomy and by many other developments.

All these matters have come to light only quite recently (though Levi-Strauss has been around for quite some time) and are not known to philosophers of science. Of course, the material may be interpreted in various ways – but the point is that it is no longer possible to take the supremacy of 'science' (whatever *that* is) for granted or to be satisfied with the bland statements T makes in his review.

H writes: 'Confronted by a world with devastating poverty, crippling social ills in the midst of plenty and political threats to the future of existence, Feyerabend is one academic who finds in ideas above all a means to hedonistic self-fulfillment.' The point would be well taken if the philosophers of science and methodologists he defends were either interested in social problems, or capable of contributing to their solution. Even the most superficial look at the grant lists of the National Science Foundation and of similar tax-supported institutions shows that such is not the case. Philosophers of science not only waste millions of valuable tax money on ridiculous projects, they get up in arms when responsible citizens suggest public supervision of their teaching and their research and especially of their use of tax money (the Baumann amendment!). Moreover, they show only contempt for forms of life different from their own. They do not examine such forms of life except in the most super-ficial manner (cf. the comments made by Kulka and Tibbets in their reviews and my reply to Tibbets above), they barely give them a cursory glance, they do not care how important they are for the lives of people (Indians, for example – *American* Indians, that is), they reject them as useless nonsense and do their best to have them eliminated. Much of the spiritual misery of the remnants of non-Western cultures in the United States is due to this uninformed intellectual fascism of most of our leading philosophers, scientists, philosophers of science. H's accusation of 'methodological onanism' should be seen in this light. H *seems* to assume that whoever rejects the ideas of modern scientific philosophers and refuses to have intercourse with them is bound to lead a lonely life. But let me assure him that onanism is not the only alternative to sleeping with *him*. There are the older philosophers and scientists, starting with the divine Plato, and there are numerous non-Western forms of life. For me the preservation of old traditions is much more important than the charades of our hyper-modern intellectuals. First, because people have a right to live as they see fit, and secondly, because non-Western traditions have solutions to 'crippling social ills and political threats. . . .' (see above, reply to T). The point of *Against Method* was that those who want to turn a narrow scientific philosophy into a public malaise without any supervision by the public have not a leg to stand on – judged by *intellectual* standards they are just one superstition among many (cf. *AM*, Chapter 18).

H is not too happy over the fact that I prefer Mach to him and his fellow philosophers. But if I do, the reason is that they have nothing worth 'drooling' about – to use his own picturesque way of speaking. Mach criticized the science of his time, both for intellectual and for social reasons, he deplored the separation of science and philosophy, he invented a new form of research that contained them both and examined the most fundamental assumptions and he thereby provided instruments for the revolution in relativity and the quantum theory. He and his followers indeed form an 'intellectual hit parade' compared with which their late successors in the 20th century are a dreary bunch. Just take H's own attempt to overcome the Hempel-Popper view of theories. It is quite amusing to see him trying to knock about absurd views, but his research is just as relevant to the 'big questions of the time' as the fight of a couple of drunks at a boxing match: the real action is in the ring – and he is not even close to it. He is not even close to these outside fights, as can be seen from his short comments on me. I say that forms of life may be deductively disjoint, he infers that I hold the Hempel-Popper view. But classes of statements, actions, attitudes can be deductively disjoint without being parts of deductive systems as he might have realized had he chosen to read Chapter 17 of *Against Method*.

H is also upset by my frequent changes of point of view. Well, thinking is a difficult business and I have not yet found the secret of at once penetrating to the very core of truth. Has he? I admit that it takes a 'court jester' to discover and pronounce such facts, for most people are unwilling and too afraid to criticize their emperors who, after all, are the source both of their intellectual life and of their income. But I wish H would not attach me to a 'Popperian court': first, because Popperianism is not a court but at most a tiny outhouse, and secondly, because I would hope for a more interesting audience than that. I am somewhat amused to read that he wants me to get off 'the public platform' – has he lost his sense of reality to such an extent that he believes his own racket to be a matter of public concern? Nor do I quite understand his remark that 'nearly everybody' now ignores me – if that is his attitude, then why did his distinguished journal publish *three* reviews of my book instead of one? Why did he publish any review at all? Admitted – the reviews were not very competent, they were barely literate but that was not the intention of of the editor as his reply shows.

The most curious suggestion comes at the end: there should be a journal for philosophers who just want to please themselves. Well, in a way such journals already exist, not one, but many. Almost every journal in the philosophy of science deals with problems that are of no interest to anyone except to a small gang of autistic intellectuals. Why do these journals not suffice? Because their contributors take their intellectual games seriously So, poor things, they have the worst of all possible worlds. They are neither 'relevant' nor do they have fun. Small wonder they are upset at somebody who does.

Chapter 5

Life at the LSE?

'"Method" lives!' exclaims John Worral at the end of his review.[1] That is quite possible – but let us see what kind of life it is!

Worral begins by saying that while there are bound to be inconsistencies in my book there are actually none. Now, if that is so then why did he bring the matter up? Why does he say that it 'would . . . be rather easy to score debating points by exhibiting inconsistencies in Feyerabend's exposition' and yet add that 'the spirit can usually be preserved quite easily by local patching'? What he means is obviously that the book is full of contradictions, that he is astute enough to have noticed them but also generous enough to separate presentation from substance and to admit that the latter may be, and perhaps is, consistent. Unfortunately this attempt at sagacious competence is quite misdirected. When earlier reviewers thought they perceived a 'contradiction' then this was due to the fact that they mistook *reductiones ad absurdum* for direct arguments (those premises are believed, defended, asserted by the author): they knew even less about arguments than Aristotle in his *Topics*. Worral is too cagey to identify the inconsistencies he says 'may' be contained in the 'letter' of my book but I have the suspicion he makes a similar mistake.[2] But – and with this we come to a much more important point: what is

---

[1] *Erkenntnis*, 1977, pp. 243–97.

[2] There is another insinuation in the first few pages and it is this: 'It must be extremely difficult' writes Worral 'to remind oneself that one's basic position gives one no right to assert any thesis positively, no right to assert that one's position is better than another, nor even any right to claim any rational cogency for one's arguments'. What Worral means to say is clearly that I do some of the things I have no right to do, but he does not dare to say so outright for fear of being made mincemeat of. But when I offer an argument I offer it to rationalists who say they will listen to arguments only and who are bound to accept it as long as it is valid in *their* terms. And the remark that I cannot assert anything positively only shows that Worral has not grasped the difference between scepticism and epistemological anarchism: it is the *sceptic* who cannot assert things positively; the anarchist can assert anything he wants and often will assert absurd things in the hope that this will lead to new forms of life.

wrong with inconsistencies? Every critic, Worral included, seems to believe that finding a contradiction in a book or in a theory reveals a fatal shortcoming. This is the first thing reviewers are looking for and one can understand why: even an ignoramus can identify inconsistencies and so has the power to reject the most beautiful theories, the most ingenious constructions of the mind. What is the force behind this argument that makes a giant-killer out of a mere logician? The force is, so the logicians tell us, that *if we apply the rules of formal logic to a contradiction* then we must assert every statement. The italicized clause shows that the argument is about the wrong object. Scientific theories, containing contradictions, progress, lead to new discoveries, expand our horizon. This means, of course, that contradictions in science are not handled according to the naive rules of formal logic – which is a criticism of logic, not of science: logical rules are too simpleminded to be able to reflect the complex structures and movements of scientific change. Of course, one can try to make science conform to such rules, but the result is frequently a loss of fruitfulness and progress (example: transition from the older quantum theory to von Neumann's account of elementary quantum mechanics[3]). Another possibility is to admit that scientific theories and informal mathematics may be inconsistent but to deny that they are 'rationally acceptable' (this is what Worral does later on in his review). The reply to this move is: so what? Inconsistent science progresses, is fruitful, but not 'rationally acceptable'. 'Rationally acceptable' science is clumsy, retards progress, cannot be easily connected with its test base. Take your pick John Worral! And don't forget that canons of rationality were originally introduced not for name calling but in the hope they would further knowledge. The hope was mistaken in the case of some canons. What is more 'rational'? To revise the canons (as one revised the canon of certainty) or to say that the failure of theories to conform to them 'does not support the idea that [the] theories are rationally acceptable'? Again, take your pick, John.

My position 'is, then, consistent' but, so Worral continues, it is 'extremely unattractive'. For example, Worral finds it unattractive 'that there should be increased state intervention in science and that parents should have the right, if they wish, to insist that their children be taught voodoo instead of science in schools'. But don't people have the right to arrange their lives as they see fit, for example in accordance with the

---

[3] Most formalizations only move the mess from one point to another. Von Neumann can now prove spectral decomposition of functions in a very orderly way – but the relation to experiment has become more chaotic than ever. Cf. *AM*, Chapter 5, n. 23.

traditions of their forefathers? And has not the revival of such traditions often shown their superiority in domains in which science makes definite claims (acupuncture; Taoism as a philosophy of science and a social philosophy etc. etc.)? Worral assumes that science is better than everything else both as a religion and as a practical device. But has he examined the matter? For example, has he compared scientific theories of cancer treatment with herbal theories? I do not think he has. And yet he wants scientists to have the right to determine what should and what should not be taught in our schools. Nor does he give a correct account of the kind of state intervention I recommend. As he presents the matter a peaceful crowd of quiet and self-paid researchers is to be rudely disturbed by an unscientific Gestapo. But the trouble is that the Gestapo tactics are to be found on the other side. What I am talking about is the extent to which scientists and scientific organizations using all available pressure tactics short of murder have succeeded in determining what has to be done with the young (education programmes which try to weed out non-scientific traditions replacing them by silly inventions such as the 'new mathematics') with older people who have problems (psychiatry; prison reform) and with the millions of tax money which they demand with the same impudence with which the church once demanded the tithe.[4] It is the duty of the state to *protect* the citizens from these chauvinistic parasites who live off the minds and the pockets of the common people – and, having to protect the state is of course also obliged to interfere. However, I quite understand why those who profit from the status quo, why scientists and their lackeys, the philosophers of science should find my suggestions 'unattractive'.

Next Worral comes to my case studies which he regards as 'the central argument of the book'. To defuse the Galileo case he produces rules that are not touched by it. I at once admit the existence of such rules. A case study is not supposed to eliminate *all* rules, but only some and so it is easy to invent other rules to which it conforms. I even do this myself. I not only show what rules are violated by a given case, I also try to show what rules were used and why they were successful. What I do assert is that for every rule and every standard a case can be constructed that goes against the rule or the standard. This is a 'bold conjecture', beloved by Popperians. I never tried to *prove* the conjecture. But I tried to make it *plausible* by presenting cases which violate basic rules and standards of

---

[4] For the Gestapo tactics used by scientific institutions cf. now Robert Jungk, *Der Atomstaat*, Munich 1977.

rationality. It is obvious that a case study that upsets a rule or a standard may not upset a weaker substitute. But it is equally obvious that in the process of weakening his rules the rationalist will eventually come very close to my own position. Lakatos is an example. Using his rationality his position is now indistinguishable from mine. Worral does not seem to have learned the lesson.

This is shown by some of his observations. For example, Worral points out that short term violations of standards do not imply their uselessness. Thus we may occasionally use ad hoc hypotheses but this does not mean that the standard or rule that enjoins us to replace them *eventually* by content increasing measures must be given up.

There are two remarks that can be made about this observation. First it admits that ad hoc hypotheses may be used. This is quite a concession for Popperians to make and it took them a long time to make it. So let us not forget that what Worral offers as such an obvious new rule is the result of an adaptive move within the Popperian circle.

Secondly, even this much more 'modest' rule is not 'immune' to criticism. I did not give the criticism in my book for I obviously could not deal with every damn rule philosophers might dream up; but I asserted, again in the form of a conjecture, that a criticism exists for every rule. The reason is that using rules and/or standards has cosmological implications.[5] For example, the rule of content increase will eventually run idle in a finite world. Research in accordance with contrary rules may help us to pinpoint such limitations and thereby to criticize the rule. This disposes of Worral's objections to the 'central argument' of my book.

Next Worral considers 'more specific points'. He tries to defuse some of them by making them sound trivial. He overlooks that I did not only address advanced thinkers of his calibre but also inductivists, naive falsificationists, Newtonians and all sorts of other people. Thus he denies that the Brownian motion case is a 'revolutionary challenge to empirist orthodoxy'. Apparently he does not know how many people still hold on to Newton's rule iv. And there are even more who are aghast at the idea that a theory can be upheld despite clear and unambiguous evidence to the contrary. Of course, none of this is found among the two or three pupils of Lakatos who still remember his teaching – but he is quite mistaken when believing that everybody has now adopted their philosophy. Then Worral gives an account of the tower argument (*not* the tower

---

[5] Philosophers of the Vienna Circle and Popperians are fond of turning cosmological principles such as the principle of causality into formal rules. As a result they eliminate circumstances that might endanger the rules.

'experiment' as he calls it) that makes it seem less 'mysterious' than my own. I agree that this is a possible account and that it was held by some of Galileo's contemporaries. But it was not held by all and it was not held by the common people *as Galileo himself says*. Had Galileo had to deal with those accepting Worral's interpretation then his problems would have been much smaller. As usual Worral considers possibilities, chooses the one that seems simple to him and then assumes that history was equally reasonable. Worral is also quite displeased (and I can understand why) at my description of Galileo as a shrewd propagandist. Now propaganda can be understood in two different ways viz. (1) as consisting of 'external' (in the sense of Lakatos) moves in favour of a theory or a research programme that conflict with 'internal' standards and (2) as consisting of misleading accounts which suppress difficulties in order to create a better press for some theory. I think I have shown that Galileo used and had to use 'propaganda' in the sense of (1) if we choose the usual 'internal' standards (up to and including Lakatos). Of course, if we choose different kinds of standards, for example if we permit standards to change, in an 'opportunistic' manner[6] from one case to the next, then the 'propaganda' turns into reason. Galileo also uses propaganda of type (2) – he gives a misleading account of the difficulties of celestial observation. Propaganda of type (2) played some role in his success and was therefore not entirely unnecessary.

And with this we come to the monster incommensurability. Needless to say, Worral does not like anything about Chapter 17. He does not like my example – Homer vs. the Presocratics – because he misunderstands the use it makes of works of art. 'As a "rationalist methodologist"' he writes 'I would find it disconcerting if it turned out that any two rival theories are, despite appearances, necessarily incommensurable; but I am not at all disconcerted or even surprised to learn that styles of drawing and painting may be incommensurable'. But my point is not that the *paintings* are mutually incommensurable but that the *cosmologies* I infer from them (and from the literature, philosophy, theology, even geography of the time) are incommensurable. Next he points out that incommensurability is unnecessary because he can easily imagine other interpretations. For example, he can easily imagine that Einsteinian mechanics contradicts rather than suspends Newtonian assumptions. But the question is not what a person who looks at the theory from some distance can imagine, the question is how his imagination fits in with ideas

---

[6] For 'opportunistic' cf. *AM*, Introduction, n. 6.

physicists find plausible, how it aids them in the interpretation of particular experiments and in understanding the relation between theory and observation. The quantum theory from the very beginning was subjected to a variety of interpretations, some of which made it incommensurable with classical physics. The debate about these interpretations was complex and is not finished, not even today. The arguments in favour of incommensurability, which one finds in the Copenhagen interpretation are quite subtle and considerably restrict the domain of what philosophers might want to imagine.[7] We must turn to these arguments when debating incommensurability and not to the naive models about theory-comparison which philosophers have developed. The same applies to relativity. Worral writes with conviction: 'Surely Einsteinian mechanics entails that a body's shape is a function of its velocity and this simply *contradicts* (his emphasis) the Newtonian assumption.' *Not* if Einstein's theory is built up in the manner of Marzke and Wheeler, i.e. without any help from classical concepts. Now it is true that in *AM* I have neither given arguments for the Copenhagen interpretation, nor discussed the interpretation of Marzke–Wheeler and its advantages. I did this in earlier papers to which I referred in footnotes and I assumed that the reader would look for the arguments there. In *AM* I presented a general model and I explained it with the help of a non scientific example. So, dear John, refute Bohr (as presented by me), refute Marzke and Wheeler or show that neither entails incommensurability, and then we shall talk again.

Worral concludes with a brief account of what I say about content increase. He admits that there are cases where loss of content occurs but asserts that such cases are very rare. Many people at the LSE, in and outside the philosophy department, seem to share this opinion. To refute it one would have to present long lists of cases where loss of content does occur. I did not do this but quoted a few paradigmatic cases that might enable the reader to do the remaining research himself. The cases I used are (1) the transition from the demon theory of mental illness to a purely behavioural account and (2) the transition from 19th century electrodynamics to the electrodynamics of relativity. The demon theory not only explains but also describes mental illness in terms of possession by demons. Statements about demons, their complex relations to each other and to their victims belong to the content of the theory. So do statements about human behaviour. During the transition statements of the first kind are dropped from the content of psychiatric (psychological) theory

---

[7] Cf. my account in *Philosophy of Science* 1968/69 (two parts).

without being replaced by other statements. The content of psychiatry (psychology) experiences a considerable shrinkage. The same is true in the case of electrodynamics. 19th century electrodynamics contains statements about properties of the ether, i.e. both about its overall properties and about its special behaviour at particular space-time regions. All these statements disappear on transition to the theory of relativity (in addition the whole theory of solid objects disappears). They are not replaced by other statements. Again there is considerable loss of content. As against these examples Worral has pointed out, in a 'Position paper' on what he calls 'Critical Rationalism' (which is simply the collection of ideas assembled by Popper and his followers) that content deals with observation statements only and that the cases I mention concern theoretical statements. This is both untrue and disingenious. It is untrue, because many statements about demons and ether properties were observable, even directly observable (as an example cf. Lodge's method of measuring ether movements or the reports of many women that the devil had an icecold member). And it is disingenious because of its relapse into old fashioned positivism. For decades Popperians have made a big fuss about the essentially theoretical nature of all statements, now that one draws consequences from their assertions they fall back upon a naive observationalist philosophy: doesn't Worral remember that the distinction between observational statements and theoretical statements depends on the theories used and that speaking of observations absolutely, as he does in his objection, means returning to old fashioned positivism? I don't have any objection to such a return, after all, old fashioned positivism was a nice theory, but I wish he would be more straightforward in his moves and admit that the condition of content increase can be upheld only by giving up the idea that all observations are shot through with theory. On the other hand, when Worral rejected Aristotle with the remark that Aristotle, while perhaps talking about things that are no longer treated by science did not give a scientific explanation of them then he is just engaged in word mongering. Scientific or not: Aristotle dealt with a wide variety of phenomena on the basis of a few simple notions. He had a theory that covered inanimate nature, animate nature, man, man's products such as his science, his philosophy, his theology (for example, he had a well worked out poetry which said, in effect, that poetry is more philosophical than history because it explains while history only describes), his theory also dealt with God and his relation to the world, it was used long after the scientific revolution in a variety of fields and with great success (Harvey, for example, was a committed Aristotelian). It is true that most British scientists rejected Aristotle because of his failure in astronomy which only showed their illiteracy in matters outside this field. It is the late after effect of this illiteracy that is used today to push aside Aristotle. Popper did it in his *Open Society* and now Worral repeats the charge, no doubt without having read a single line of Aristotle. This, incidentally, is a rather widespread feature of 'progress': some people move ahead a few inches in a narrow field. They assume that their 'progress' covers a much wider area. They support this assumption with an account of the opposition that is heavily biased and shows a great deal of ignorance. The ignorance is soon regarded as knowledge and handed on, authoritatively, from teacher to pupil to grandpupil. And so intellectual midgets can pose as giants and give the impression that they have superseded the real giants of the past. Critical rationalism is one of the schools that owes its fame to this phenomenon.

Summarizing the criticism – what do we see a methodologist engaged in? We see him engaged in weakening methodological rules whenever someone else has shown the stronger version to be in conflict with scientific practice; he criticizes historical examples on the basis of what *he* would have thought on the matter, he opposes the results of complex arguments (which he does not know) by saying that 'surely' a different account is also possible, he upholds unrealistic theories of rationality and scientific change by refusing to describe achievements in their terms (contradictory theories may be fruitful, but they are not 'rational') and he generally engages in word-mongering of all sorts (Aristotle not 'scientific'). Considering that there are many people engaged in these interesting activities we must admit that methodology is still very much alive, even at the LSE; but it is not the kind of life a reasonable person would want to live.

Index